Grades 4-8

CHECK YOUR WORK

Written by

Paul Swan

Other titles by Paul Swan

Number Grids
Grades 3-8 2-163

Magic with Math
Grades 5-8 2-162

Patterns in Mathematics
Grades 3-6 2-165

Mastergrids for Mathematics
Elementary 2-5195

Messy Math
Grades 4-7 2-161

Published with the permission of R.I.C. Publications Pty. Ltd.

First published by R.I.C. Publications Pty. Ltd., Perth, Western Australia. Revised by Didax Educational Resources.

Printed in the United States of America.

Order Number 2-5234
ISBN 1-58324-198-1

A B C D E F 09 08 07 06 05

395 Main Street
Rowley, MA 01969
www.didax.com

Foreword

In just about every computation lesson, teachers can be heard telling students to "check their work." Generally this statement is made in response to a student who has written down a silly answer or has accepted a result that can't possibly be correct. It is of concern to teachers that students rarely question the results of a calculation or the validity of the result.

It has also become increasingly vital in this calculator and computer age that students be taught how to check their work, but when we ask students to check their work they generally:

- *turn to the answer page,*
- *reach for a calculator, or*
- *don't bother because it is too much work.*

Check your Work has been designed to address the question of how to check the results of a computation. In this book, you will find a range of options for checking the results of a calculation, beginning with simple, rough methods and progressing to more sophisticated approaches. The simple "calculation checks" suggested are intuitive and rely on students developing a feel for number or "number sense." These methods are ideal for children who do not like checking their work because of the effort involved.

The more sophisticated methods rely on students' "understanding" of basic properties of number, such as odd and even. The most sophisticated methods are designed to give students an alternative approach to a calculation that will allow them to verify its results.

While many of the techniques suggested in this book may be used to check all types of calculation, specific reference is made to calculator-based computations. Calculators are now commonly used in schools, so specific issues associated with using a calculator are also addressed, along with the provision of simple ways of checking the results of a calculation.

Contents

Teacher's notes

Check your work contains seven units. Each unit is described in detail in the following pages.

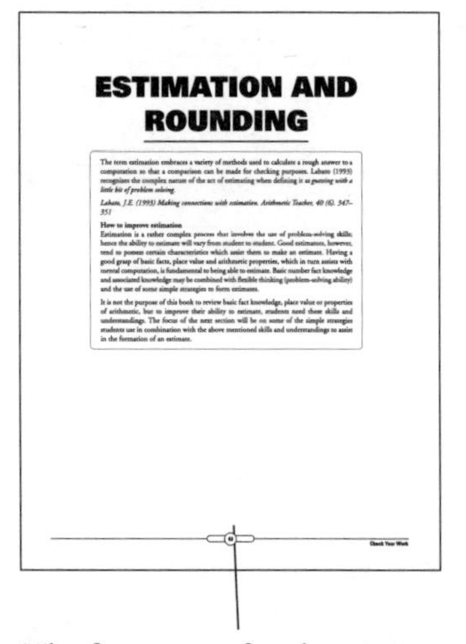
ESTIMATION AND ROUNDING

The answers for activity questions are provided.

Number intuition

Odd and even patterns – 1

Answers

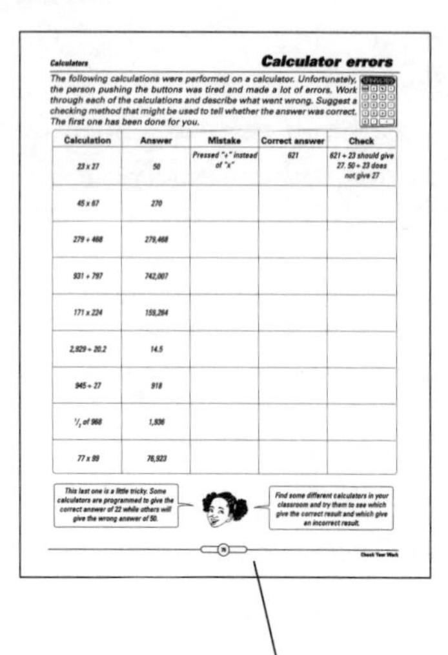
Calculators

Calculator errors

Calculation	Answer	Mistake	Correct answer	Check
23 x 27	50	Pressed "+" instead of "x"	621	621 ÷ 23 should give 27. 50 ÷ 23 does not give 27
45 x 67	270			
279 + 468	279,468			
931 + 797	742,007			
171 x 224	159,284			
2,929 ÷ 20.2	14.5			
945 ÷ 27	918			
1/2 of 968	1,936			
77 x 99	76,923			

The first page of each unit is a cover page. The content of the unit is summarized here.

Teacher information is included to provide additional information and examples for teachers and students.

Activities requiring calculators.

A variety of student pages are provided. They may take the form of questions and answers, group activities, calculator activities, or games.

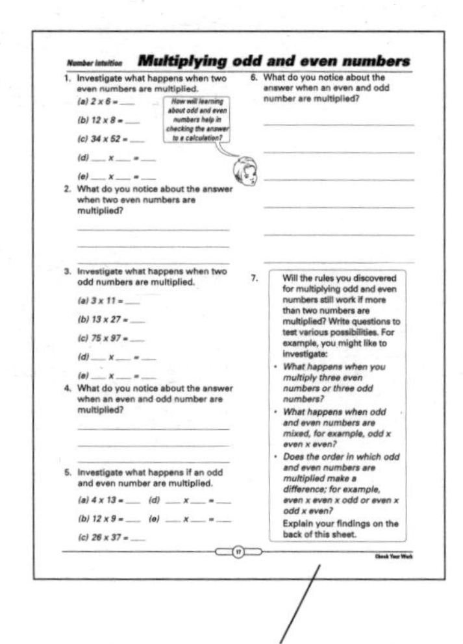
Number intuition

Multiplying odd and even numbers

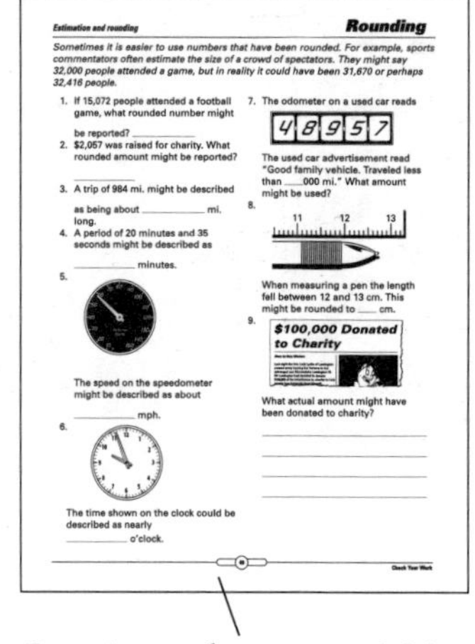
Estimation and rounding

Rounding

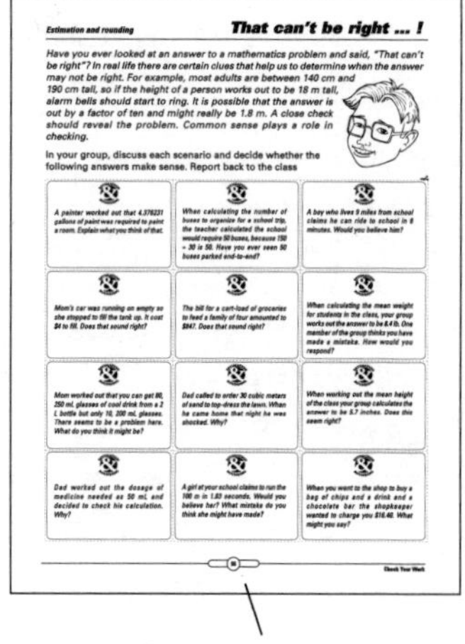
Estimation and rounding

That can't be right ... !

Suggestions on how to complete the investigations are also included.

Question and answer activities.

Group activities.

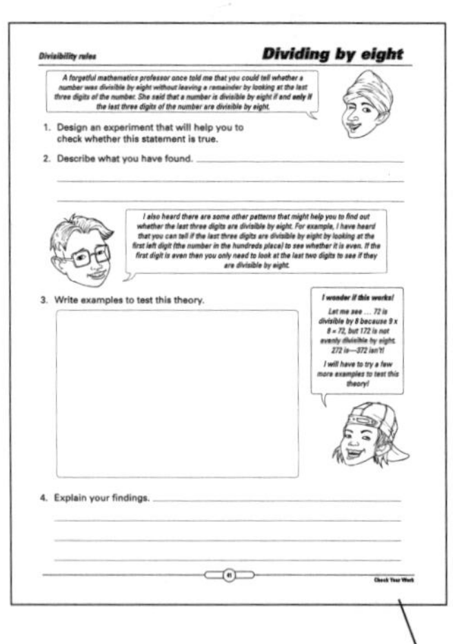
Divisibility rules

Dividing by eight

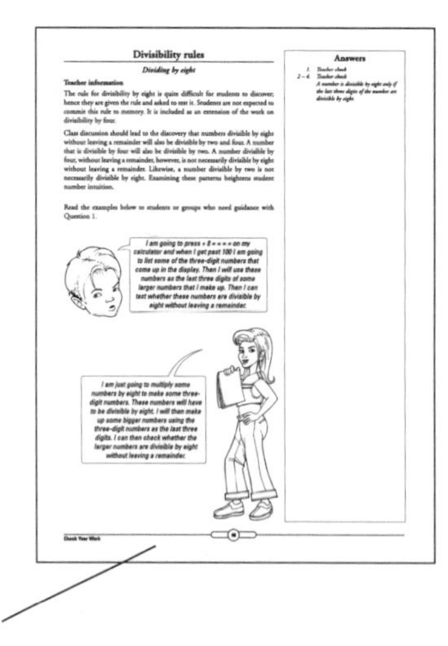
Divisibility rules

Dividing by eight

Answers

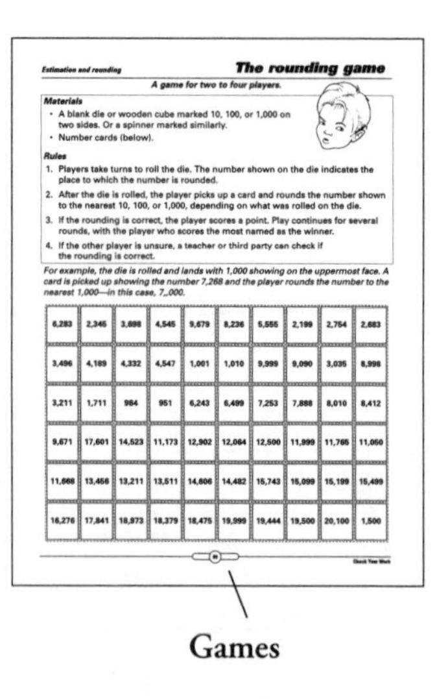
Estimation and rounding

The rounding game

A game for two to four players.

6,283	2,345	3,888	4,545	9,679	8,236	5,555	2,199	2,754	2,683
3,496	4,169	4,332	4,547	1,001	1,010	9,999	9,090	3,035	8,998
3,211	1,711	984	951	6,243	6,489	7,253	7,888	8,010	8,412
9,671	17,601	14,523	11,173	12,902	12,064	12,500	11,999	11,785	11,050
11,868	13,456	13,211	13,511	14,606	14,482	15,743	15,099	15,199	15,499
16,276	17,841	18,972	18,379	18,475	19,999	19,444	19,500	20,100	1,500

Investigations

Games

Teacher's notes

Also included:

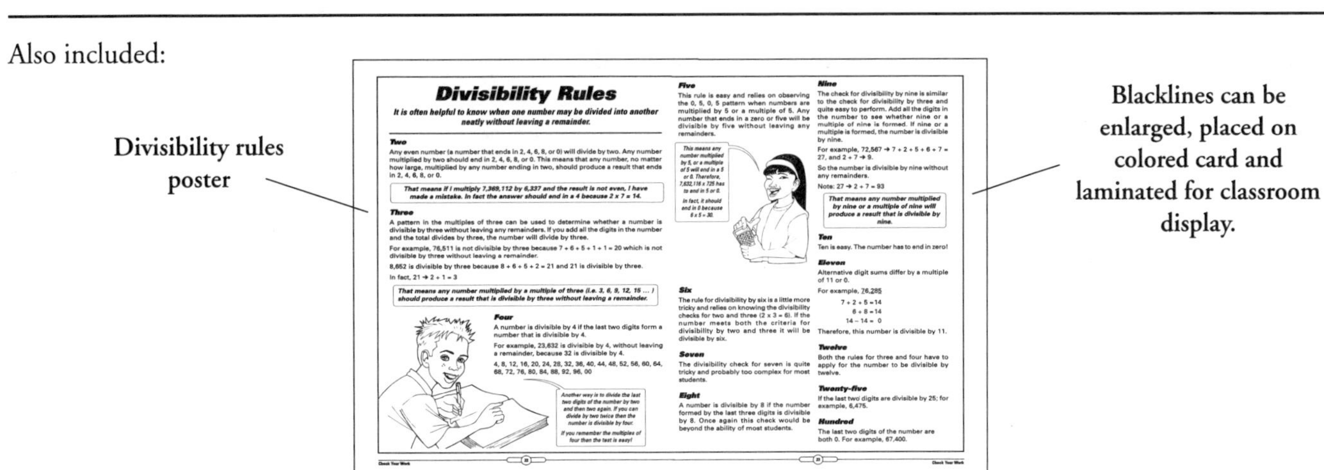

Divisibility Rules

It is often helpful to know when one number may be divided into another neatly without leaving a remainder.

Two

Any even number (a number that ends in 2, 4, 6, 8, or 0) will divide by two. Any number multiplied by two should end in 2, 4, 6, 8, or 0. This means that any number, no matter how large, multiplied by any number ending in two, should produce a result that ends in 2, 4, 6, 8, or 0.

That means if I multiply 7,369,112 by 6,337 and the result is not even, I have made a mistake. In fact the answer should end in a 4 because 2 x 7 = 14.

Three

A pattern in the multiples of three can be used to determine whether a number is divisible by three without leaving any remainders. If you add all the digits in the number and the total divides by three, the number will divide by three.

For example, 76,511 is not divisible by three because 7 + 6 + 5 + 1 + 1 = 20 which is not divisible by three without leaving a remainder.

8,652 is divisible by three because 8 + 6 + 5 + 2 = 21 and 21 is divisible by three.

In fact, 21 → 2 + 1 = 3

That means any number multiplied by a multiple of three (i.e. 3, 6, 9, 12, 15 ...) should produce a result that is divisible by three without leaving a remainder.

Four

A number is divisible by 4 if the last two digits form a number that is divisible by 4.

For example, 23,632 is divisible by 4, without leaving a remainder, because 32 is divisible by 4.

4, 8, 12, 16, 20, 24, 28, 32, 36, 40, 44, 48, 52, 56, 60, 64, 68, 72, 76, 80, 84, 88, 92, 96, 00

Five

This rule is easy and relies on observing the 0, 5, 0, 5 pattern when numbers are multiplied by 5 or a multiple of 5. Any number that ends in a zero or five will be divisible by five without leaving any remainders.

Six

The rule for divisibility by six is a little more tricky and relies on knowing the divisibility checks for two and three (2 x 3 = 6). If the number meets both the criteria for divisibility by two and three it will be divisible by six.

Seven

The divisibility check for seven is quite tricky and probably too complex for most students.

Eight

A number is divisible by 8 if the number formed by the last three digits is divisible by 8. Once again this check would be beyond the ability of most students.

Nine

The check for divisibility by nine is similar to the check for divisibility by three and quite easy to perform. Add all the digits in the number to see whether nine or a multiple of nine is formed. If nine or a multiple is formed, the number is divisible by nine.

For example, 72,567 → 7 + 2 + 5 + 6 + 7 = 27, and 2 + 7 → 9.

So the number is divisible by nine without any remainders.

Note: 27 → 2 + 7 = 93

That means any number multiplied by nine or a multiple of nine will produce a result that is divisible by nine.

Ten

Ten is easy. The number has to end in zero!

Eleven

Alternative digit sums differ by a multiple of 11 or 0.

For example, 76,285

7 + 2 + 5 = 14
6 + 8 = 14
14 – 14 = 0

Therefore, this number is divisible by 11.

Twelve

Both the rules for three and four have to apply for the number to be divisible by twelve.

Twenty-five

If the last two digits are divisible by 25; for example, 6,475.

Hundred

The last two digits of the number are both 0. For example, 67,400.

Check Your Work

Check Your Work

Divisibility rules poster

Blacklines can be enlarged, placed on colored card and laminated for classroom display.

Teachers may use these review pages to assess student understanding of some of the checking techniques outlined in this book.

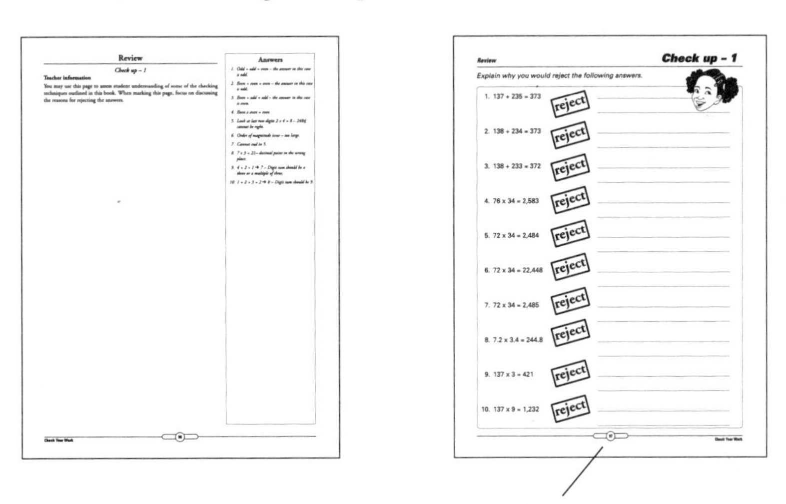

Review

Check up – 1

Teacher information

You may use this page to assess student understanding of some of the checking techniques outlined in this book. When marking this page, focus on discussing the reasons for rejecting the answers.

Answers

Check Your Work

Review

Check up – 1

Explain why you would reject the following answers.

1. 137 + 235 = 373 reject
2. 138 + 234 = 373 reject
3. 138 + 233 = 372 reject
4. 76 x 34 = 2,583 reject
5. 72 x 34 = 2,484 reject
6. 72 x 34 = 22,448 reject
7. 72 x 34 = 2,485 reject
8. 7.2 x 3.4 = 244.8 reject
9. 137 x 3 = 421 reject
10. 137 x 9 = 1,232 reject

Check Your Work

Students must give reasons for why the answers are rejected.

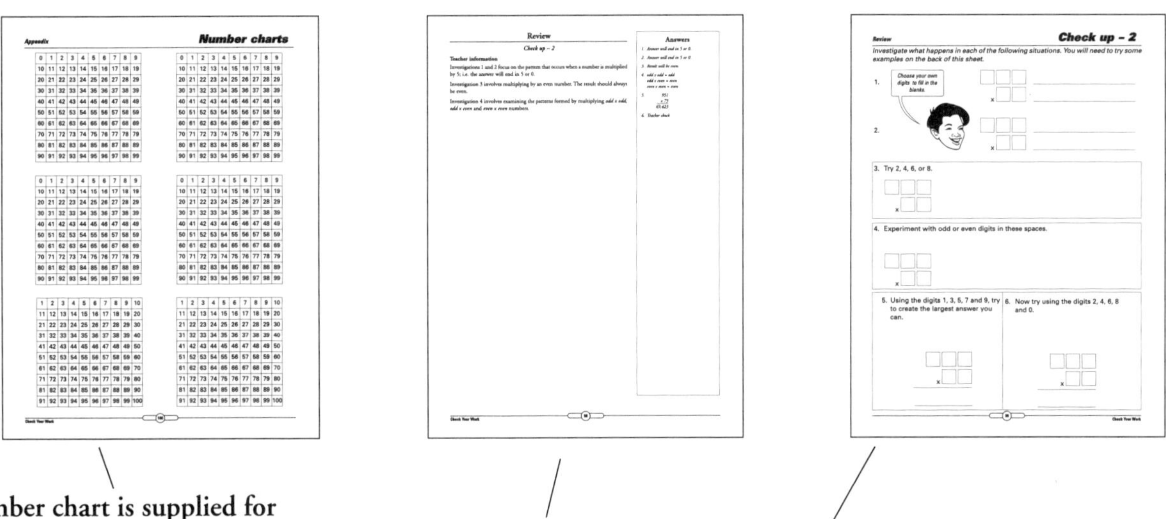

Appendix

Number charts

Check Your Work

Review

Check up – 2

Teacher information

Answers

Check Your Work

Review

Check up – 2

Investigate what happens in each of the following situations. You will need to try some examples on the back of this sheet.

1.

2.

3. Try 2, 4, 6, or 8.

4. Experiment with odd or even digits in these spaces.

5. Using the digits 1, 3, 5, 7 and 9, try to create the largest answer you can.

6. Now try using the digits 2, 4, 6, 8 and 0.

Check Your Work

A number chart is supplied for teacher and student use.

Students use their knowledge of patterns in the multiplication tables and odd and even number patterns to investigate problems.

Teacher's notes

When students are asked to check the work they invariably use one of three options:

- *Don't bother because it is too much work,*
- *Reach for a calculator, or*
- *Turn to the answer page.*

This book is designed to broaden the range of checking strategies for students. It begins with very simple—almost intuitive—forms of checking, such as whether the answer should be odd or even, to more formal methods of checking such as casting out nines. Some of the *calculation checks* suggested in this book are simple or intuitive and rely on developing a *feel for number* or *number sense.* These methods are ideal for children who do not like checking their work because of the effort that is required. In order for students' intuition about number, or number sense, to grow, they must develop an awareness of various number patterns and properties of number. Many of the activities contained in this book are designed to increase students' number sense and thereby improve their judgment about the reasonableness of a result. There are several methods students might employ when checking the results of a computation.

These include:

using an intuitive form of checking

- *considering the odd and evenness of results*
- *adding two odd numbers only gives an even result*
- *making use of number patterns*
- *multiplying a number by 5 means that the result must end in 5 or 0*
- *divisibility rules*
- *making use of common sense*

making use of an estimate

- *order of magnitude*
- *front-end methods*
- *using rounding*
- *27 x 38 is about 30 x 40 (near doubles)*

casting out nines

repeating the calculation

- *repeat it carefully in the same way*
- *repeat it again using a different method*
- *using a different operation*
- *using a different form of calculation*

Calculators are often employed as checking devices, but one might ask, "If they are so readily available, why weren't they used to complete the calculation in the first place?" All the methods described in this publication may be used to check calculations done with pencil and paper, or with a calculator, and many of the methods may be used to monitor the results of mental calculation.

Overview of checking methods

The checking methods in this book may be grouped under three broad headings.

- *Basic*
- *Intermediate*
- *Advanced (Sophisticated)*

This book begins with some basic checking methods and moves through intermediate to advanced checking methods. Not all methods will be applicable to all situations and some students will prefer some methods to others. You are encouraged to discuss the various facets and features of each method so that students develop a repertoire of approaches for checking their work.

Basic methods

Basic methods rely on the use of common sense, or number sense. Students need to ask themselves whether the answer sounds right or whether the result is reasonable given the context. One basic technique that may be applied is to make an order of magnitude check; that is, deciding whether the answer is way too big or too small.

Intermediate check

This form of checking involves using one or more features of an answer to determine whether the calculation is incorrect. For example, knowledge of number patterns should lead a student to realize that a number multiplied by ten or a multiple of ten should end in zero. Likewise, realizing that adding two odd numbers will produce an even number means that a student only need look at the ones digit in an answer to determine whether it is incorrect.

Intermediate checks often rely on having made an estimate prior to performing the calculation. The estimate may not involve using the specific estimation technique but rather a benchmark figure. For example, when multiplying 6.2 by 5.4, there should be no need to count decimal places to determine the position of the decimal point. Rather, a student should recognize the result should be around 30 and therefore place the decimal point accordingly.

Advanced or Sophisticated checking methods

These methods are generally more time consuming than either the basic or intermediate checks. They might rely on carrying out a calculation in a different way or using an inverse operation (i.e. subtraction to check addition).

Number intuition (pages 9 to 11)

Most children avoid checking their work because of the time and effort it takes to perform calculation checks. The two simplest methods involve developing students' basic instincts. Number intuition involves using knowledge of patterns to alert students to the possibility that the answer may be wrong. These patterns include odd and even number patterns (adding two odd numbers cannot give an odd number) and rules of divisibility. For example, when multiplying by five, the result must end in 5 or 0. If a number has been multiplied by 675 then a student should expect the result to end in 0 or 5. Alarm bells should ring if the answer ends in any other number.

Closely allied to number intuition is the use of common sense. The use of common sense when checking the results of a calculation is often associated with context. When calculations are given in context, certain contextual clues can alert the student to possible errors. For example, when asked to calculate the number of buses required to transport 150 students if each bus holds 30, many students answer 50; but when asked if they have ever seen 50 buses end to end, most realize their mistake.

Estimation (pages 43 to 69)

Students are often encouraged to estimate prior to performing a calculation to compare the calculated result to the estimated result. Students should be encouraged to follow a three-pronged approach: Estimate, Calculate, Evaluate. Unfortunately, most students do not really understand how to estimate. Some calculate the exact result and then "adjust" it to look like an estimate. This technique defeats the purpose of making an estimate, is tedious and is soon generally abandoned. Most students are taught to round when forming an estimate, but some become confused when rounding. Once students master the process of rounding, estimation becomes relatively simple. A simpler but less used approach involves using the front-end method, which avoids the need to learn rounding skills. Less sophisticated estimation techniques involve estimating the order of magnitude, such as whether the answer is likely to be in the hundreds, thousands, tens of thousands and so on. Students may also estimate a range of possible answers; e.g. the answer will lie between 100 and 400.

Repeating a calculation (pages 71 to 75)

When using paper-and-pencil methods, most students do not like repeating a calculation but in the age of the calculator very little effort is required to repeat a calculation to check the result. Repeating a calculation can produce a dilemma, because if the second result does not match the first, a decision has to be made as to which result might be correct or the calculation must be repeated another time. Students who try this approach often simply repeat the calculation in the same way it was originally executed. There are several options that can be used.

- Repeat the calculation carefully exactly the same way.
- Repeat in a different way; e.g. use an inverse calculation; i.e. work back to find the answer. For example, 1470 ÷ 35 = 42 may be checked by multiplying 35 by 42 to see if the result is 1470.
- Repeat using a different method.

Casting out nines (pages 77 to 87)

Casting out nines is a more sophisticated checking approach that may be used to tell you if the answer is likely to be correct and tell you definitely whether the answer is wrong. At first, casting out nines takes a little time to master, but once mastered it may be executed rapidly, especially for addition and subtraction.

Teacher's notes

Responding to children who ask – "Is this right?"

Sometimes when you're busy it may be tempting to reply *yes* or *no*—and on the odd occasion it may be fine. But in the long term offering a simple *yes/no* response will breed a class of students dependant on you for confirmation of everything they do. Even though it may take time, in the long run it will be better to avoid giving *yes/no* responses. Here are some suggestions to avoid giving such a response.

- *Make better use of wait time. Don't respond right away but rather consider the question and why the child is asking it.*
- *Respond with a question. For example, you could ask "What do you think?" "Why are you asking? – Do you think it is wrong?"*
- *Ask the child how sure he/she is that the answer is correct. "On a scale of 1–10, how sure are you that the answer is correct?"*
- *Provide another example for the child to try. This is particularly effective when the answer is incorrect. For example, if a child asks if 9 x 6 is 45, you might ask what nine fives are. Alternatively, if time permits, the pattern when multiplying by nines might be examined or the relationship between the nine and ten times table might be considered.*
- *Ask "How might you tell if the answer was right?" or "What might alert you to the answer being wrong?"*
- *Work through a suitable strategy for determining whether an answer is likely to be correct.*
- *How did you do it?*
- *Show me, using materials.*
- *What does this part mean?*
- *If the student has made a basic fact error, you might ask "what are six threes?" (the specific basic fact).*
- *Suggest an appropriate checking approach.*
- *Ask – "Is that reasonable?"*

NUMBER INTUITION

Number intuition involves developing a "feel for numbers." This feeling is developed by raising students' awareness of patterns and relationships. As students develop this ability, they will make comments like "That doesn't look right" or "That doesn't sound right." Number intuition should not be confused with the broader idea of *number sense*. While *number intuition* should be part of *number sense*, the latter involves much more.

Number intuition

Odd and Even Cards

Teacher information

Some students have trouble discerning the relationships between odd and even numbers when they are presented in symbolic form. For those students who tend to understand relationships better when presented in visual form, you might try using some odd/even cards. Once cut out, these cards may be manipulated to investigate what happens when odd and even numbers are added.

For example, when adding two odd numbers, students can see that joining two odd number cards produces an even result. Students will need to experiment with various combinations of odd and even numbers to discover the relationship when adding odd and even numbers.

For example:

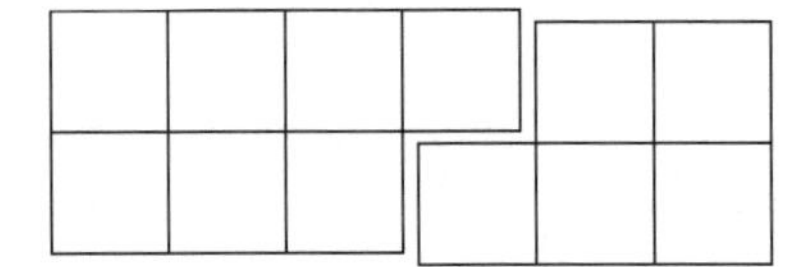

7 (odd) and 5 (odd) make 12 (even)

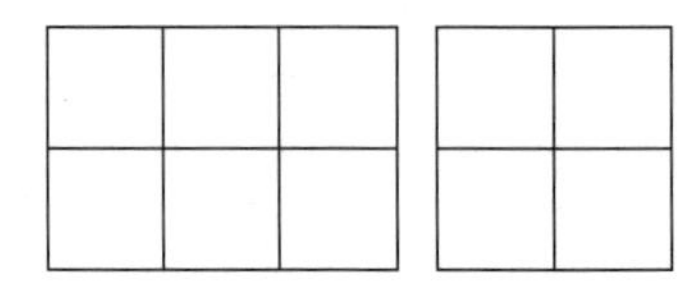

6 (even) and 4 (even) make 10 (even)

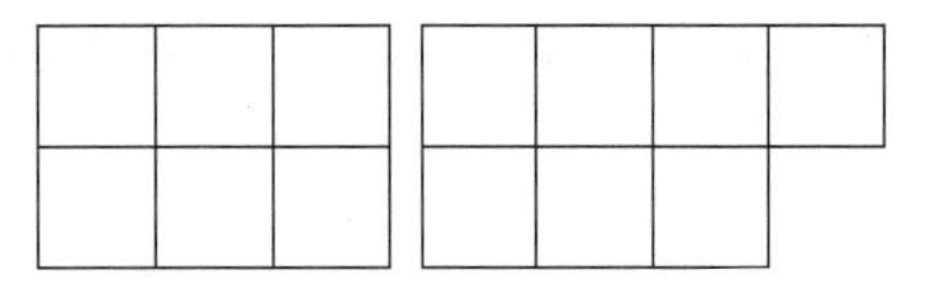

6 (even) and 7 (odd) make 13 (odd)

When examining the adding of an even number, students could be asked to explore what happens if an odd number and an even number are added.

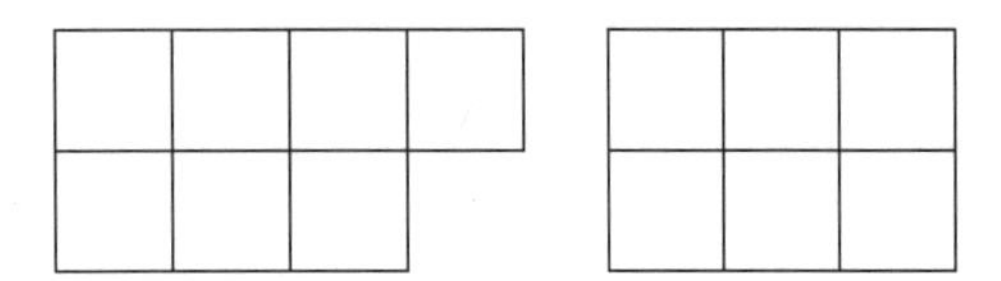

In this way the communicative property of addition is highlighted.

Answers

1. (a) 6 (even), rectangle
   (b) 10 (even), rectangle
   (c) odd, 4 (even), rectangle
   (d) odd, odd, 8 (even), rectangle
   (e) even, odd, 5 (odd), no shape
   (f) even, odd, 11 (odd) no shape
2. *Always even*
3. *Always even*
4. *Always odd*
5. *Order does not matter for addition.*

Odd and even cards

1. Cut out the odd/even cards below. Make shapes to match the number sentences. Complete the number sentences.

   (a) 2 (even) and 4 (even) make ___ (   ).

   What shape is formed?

   ____

   (b) 4 (even) and 6 (even) make ___ (   ).

   What shape is formed?

   ____

   (c) 1 (odd) and 3 (   ) make ___ (   ).

   What shape is formed?

   ____

   (d) 5 (   ) and 3 (   ) make ___ (   ).

   What shape is formed?

   ____

   (e) 4 (   ) and 1 (   ) make ___ (   ).

   What shape is formed?

   ____

   (f) 6 (   ) and 5 (   ) make ___ (   ).

   What shape is formed?

   ____

2. What do you notice when two even cards are combined?

   ____

   ____

   ____

   ____

3. What do you notice when two odd cards are combined?

   ____

   ____

   ____

   ____

4. What do you notice when an odd and an even card are combined?

   ____

   ____

   ____

   ____

5. Does it matter what order the cards are combined in?

   ____

   ____

   ____

Number intuition

Odd and even patterns – 1

Teacher information

We often ask children to consider the following patterns.

- *odd number* + *odd number* = *even number*
- *odd number* + *even number* = *odd number*
- *even number* + *even number* = *even number*
- *odd number* – *odd number* = *even number*
- *odd number* – *even number* = *odd number*
- *even number* – *odd number* = *odd number*
- *even number* – *even number* = *even number*

Rarely do children apply what they learn about these patterns as a means of checking a calculation.

These checks can be stated more simply.

- *The answer to an addition problem can only be odd if there is a mix of even and odd numbers in the question.*

How will learning about odd and even numbers help in checking a calculation? It is easy to tell whether a number is odd or even simply by looking at the ones digit. Once you are aware of the pattern, all you need to do is inspect the ones digits of the numbers that are added, and the answer, to determine whether the answer is definitely wrong or possibly correct.

Answers

1. (a) 8
(b) 86
(c) 330
(d) Teacher check
(e) Teacher check
2. When two even numbers are added, an even result is produced.
even + even = even
3. (a) 14
(b) 40
(c) 172
(d) Teacher check
(e) Teacher check
4. When two odd numbers are added, an even result is produced.
odd + odd = even
5. (a) 17
(b) 63
(c) 145
(d) Teacher check
(e) Teacher check
6. When an even and an odd number are added an odd result is produced.
even + odd = odd
7. Teacher check

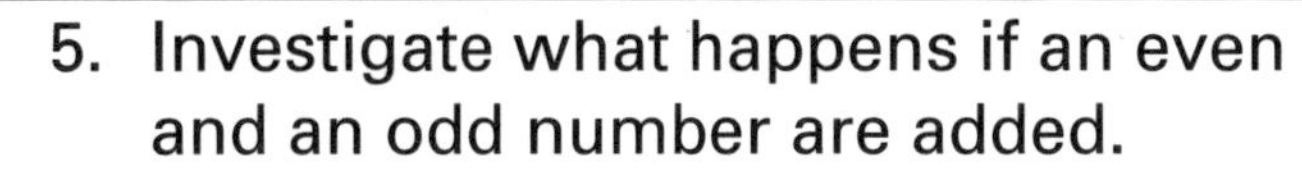

1. Investigate what happens if two even numbers are added.

   (a) 2 + 6 = ____

   (b) 34 + 52 = ____

   (c) 106 + 224 = ____

   (d) ____ + ____ = ____

   (e) ____ + ____ = ____

2. What do you notice about the answer when two even numbers are added?

   ____

   ____

   ____

   ____

3. Investigate what happens if two odd numbers are added.

   (a) 3 + 11 = ____

   (b) 13 + 27 = ____

   (c) 75 + 97 = ____

   (d) ____ + ____ = ____

   (e) ____ + ____ = ____

4. What do you notice about the answer when two odd numbers are added?

   ____

   ____

   ____

   ____

How will learning about odd and even numbers help in checking the answer to a calculation?

I wonder what happens with subtraction?

5. Investigate what happens if an even and an odd number are added.

   (a) 4 + 13 = ____

   (b) 26 + 37 = ____

   (c) 48 + 97 = ____

   (d) ____ + ____ = ____

   (e) ____ + ____ = ____

6. What do you notice about the answer when an even and an odd numbers are added?

   ____

   ____

   ____

   ____

7. Will the rules you discovered for adding odd and even numbers still work if more than two numbers are added? Write questions to test various possibilities. For example, you might like to investigate:

   - *What happens if you add three even numbers or three odd numbers?*
   - *What happens when odd and even numbers are mixed; for example, odd + even + even?*
   - *Does the order in which odd and even numbers are added, make a difference; for example even + even + odd or even + odd + even?*

   Explain your findings on the back of this sheet.

Number intuition

Odd and even patterns – 2

Teacher information

Odd and even patterns – 2 expands and consolidates previous work involving odd and even numbers. After completing this activity, students should have a clear idea of what happens when:

- *two odd numbers are added*
- *two even numbers are added*
- *an odd and an even number are added*

Students should be encouraged to apply their knowledge of these patterns when checking their work.

Answers

1. (a) *even*
   (b) *Teacher check*
   (c) *Teacher check*
   (d) *Teacher check*
2. (a) *Teacher check*
   (b) *Teacher check*
   (c) *Teacher check*
3. *odd, even*
4. (a) *odd*
   (b) *13*
   (c) *113*
   (d) *417*
5. *Teacher check*
6. (a) *Teacher check*
   (b) *Teacher check*
   (c) *Teacher check*
7. *odd, even, even*
8. (a) *correct*
   (b) *incorrect*
   (c) *incorrect*
   (d) *correct*
   (e) *correct*
   (f) *incorrect*
9. *Focus on the last digit in each number to decide if the answer will be odd or even.*

Odd and even patterns – 2

What happens when you add two odd numbers?

1. Try adding some odd numbers together and note whether the answer is even or odd.

   (a) odd + odd = ◯ *odd* ◯ *even*

   (b) ____ + ____ = ____ ◯ *odd* ◯ *even*

   (c) ____ + ____ = ____ ◯ *odd* ◯ *even*

   (d) ____ + ____ = ____ ◯ *odd* ◯ *even*

2. Choose three more pairs of odd numbers and test your ideas.

   (a) ____ + ____ = ____ ◯ *odd* ◯ *even*

   (b) ____ + ____ = ____ ◯ *odd* ◯ *even*

   (c) ____ + ____ = ____ ◯ *odd* ◯ *even*

3. Complete the following sentence.

   When you add two ________ numbers the answer is always

   ________.

What happens when you add an odd and even number?

4. Try adding some odd and even numbers together and note whether the answer is odd or even.

   (a) odd + even ◯ *odd* ◯ *even*

   (b) 7 + 6 = ____ ◯ *odd* ◯ *even*

   (c) 65 + 48 = ____ ◯ *odd* ◯ *even*

   (d) 243 + 174 = ____ ◯ *odd* ◯ *even*

5. What do you notice about the answers?

   ____

   ____

6. Choose three pairs of odd and even numbers and test your ideas.

   (a) ____ + ____ = ____ ◯ *odd* ◯ *even*

   (b) ____ + ____ = ____ ◯ *odd* ◯ *even*

   (c) ____ + ____ = ____ ◯ *odd* ◯ *even*

7. Complete the following sentence.

   When you add an ________ number and an ________ number, the answer is always ________.

8. **Without actually performing a calculation,** use what you have learned about adding odd and even numbers to decide whether the following answers are correct or incorrect.

   (a) 6 + 18 = 24 ◯ *correct* ◯ *incorrect*

   (b) 34 + 26 = 61 ◯ *correct* ◯ *incorrect*

   (c) 9 + 11 = 19 ◯ *correct* ◯ *incorrect*

   (d) 61 + 31 = 92 ◯ *correct* ◯ *incorrect*

   (e) 17 + 14 = 31 ◯ *correct* ◯ *incorrect*

   (f) 57 + 24 = 80 ◯ *correct* ◯ *incorrect*

9. Write a general rule for checking that an answer to an addition problem is correct. (Hint: Mention looking at the last digit.)

   ____

   ____

   ____

   ____

   ____

   ____

Number intuition

Multiplying odd and even numbers

Teacher information

We often ask children to consider the following patterns.

- *odd x odd = odd*
- *odd x even = even**
- *even x even = even*

* *Multiplying an odd number by an even number will produce the same result as multiplying an even number by an odd number. Students should explore both situations and discover this for themselves.*

Rarely do children apply what they learn about these patterns as a means of checking a calculation.

These checks can be stated more simply.

- *The answer to a multiplication can only be odd if two odd numbers are multiplied.*

How will learning about odd and even numbers help in checking a calculation? It is easy to tell whether a number is odd or even simply by looking at the ones digit. Once you are aware of the pattern, all you need to do is inspect the ones digits of the numbers that are multiplied, and the answer, to determine whether the answer is definitely wrong or possibly correct.

Answers

1. (a) 12
(b) 96
(c) 1768
(d) Teacher check
(e) Teacher check

2. When two even numbers are multiplied an even result is produced.
even x even = even

3. (a) 33
(b) 351
(c) 7275
(d) Teacher check
(e) Teacher check

4. When two odd numbers are multiplied an odd result is produced.
odd x odd = odd

5. (a) 52
(b) 108
(c) 962
(d) Teacher check
(e) Teacher check

6. When an even and an odd number are multiplied an even result is produced.
even x odd = even

7. Teacher check
even x even x even = even
odd x odd x odd = odd

When you multiply three numbers, all three must be odd in order for the answer to be odd.

The order in which the multiplication occurs does not matter. The result is the same.

Multiplying odd and even numbers

1. Investigate what happens when two even numbers are multiplied.

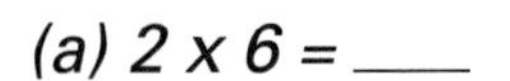

   (a) 2 x 6 = ____

   (b) 12 x 8 = ____

   (c) 34 x 52 = ____

   (d) ____ x ____ = ____

   (e) ____ x ____ = ____

How will learning about odd and even numbers help in checking the answer to a calculation?

2. What do you notice about the answer when two even numbers are multiplied?

____

____

____

3. Investigate what happens when two odd numbers are multiplied.

   (a) 3 x 11 = ____

   (b) 13 x 27 = ____

   (c) 75 x 97 = ____

   (d) ____ x ____ = ____

   (e) ____ x ____ = ____

4. What do you notice about the answer when an even and odd number are multiplied?

____

____

____

5. Investigate what happens if an odd and even number are multiplied.

   (a) 4 x 13 = ____ (d) ____ x ____ = ____

   (b) 12 x 9 = ____ (e) ____ x ____ = ____

   (c) 26 x 37 = ____

6. What do you notice about the answer when an even and odd number are multiplied?

____

____

____

____

____

____

7. Will the rules you discovered for multiplying odd and even numbers still work if more than two numbers are multiplied? Write questions to test various possibilities. For example, you might like to investigate:

   - *What happens when you multiply three even numbers or three odd numbers?*
   - *What happens when odd and even numbers are mixed, for example, odd x even x even?*
   - *Does the order in which odd and even numbers are multiplied make a difference; for example, even x even x odd or even x odd x even?*

   Explain your findings on the back of this sheet.

Number intuition

The answer should be ...

Teacher information

When using a calculator, students sometimes press too many keys. The result will either be too large or too small. Likewise when multiplying by powers of ten, students often don't include enough digits.

An order of magnitude check (i.e. is the answer likely to be in the hundreds, thousands, millions ...) should help students avoid these errors or at least alert them to potential problems. The following activities will assist students to develop a feel for the potential size of an error.

Sometimes students include too many or too few digits in the answer. Order of magnitude checks (i.e. Is the answer going to be in the hundreds, thousands, millions?) will help them avoid this type of error.

How large would you expect the product of two, two-digit numbers to be? Students may be encouraged to look for patterns which will help them make order of magnitude checks. For example, a two-digit by two-digit multiplication will produce either a three-digit or four-digit answer (i.e. an answer in the hundreds or perhaps thousands). This type of check is particularly useful when using a calculator, as it is often easy to push an extra key when calculating.

The most common examples students will come across are listed below.

- *Two-digit x two-digit ➔ three- or four-digit answer*
- *Two-digit x three-digit ➔ four- or five-digit answer*
- *Three-digit x three-digit ➔ five- or six-digit answer*
- *Three-digit x four-digit ➔ six- or seven-digit answer*

After a while the pattern becomes obvious.

Answers

1. *Teacher check*
2. *Teacher check*
3. *When adding two, two digit numbers the answer is likely to be in the tens (two digits) or the hundreds (three digits).*
4. *Teacher check*
5. *Teacher check*
6. *Students should mark (b), (d) and (e) incorrect.*
7. *Teacher check*

The answer should be ...

When you read a question, do you ever consider how big the answer will be? Try it! Consider if you think the answer will be in tens, hundreds, thousands ... or bigger!

1. How large do you think an answer will be if two, two-digit numbers are added?

tens	hundreds	thousands	bigger

2. Create some two-digit numbers and add them.

(a) 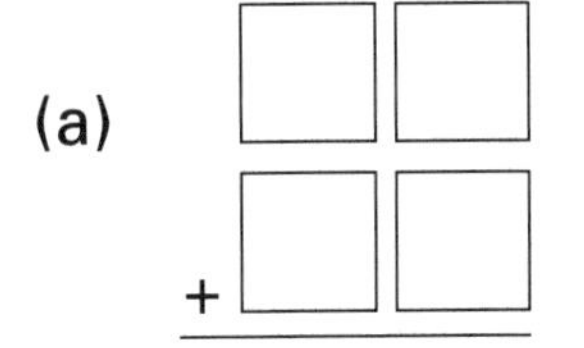

(c) 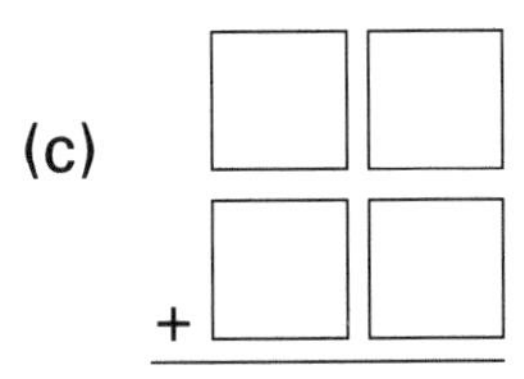

(b) 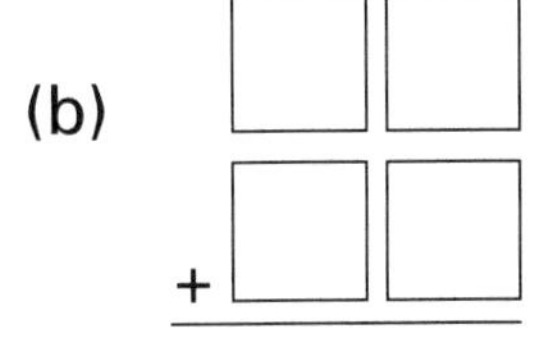

(d) 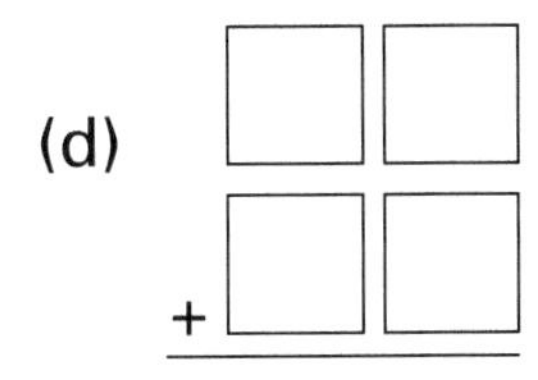

3. When adding two, two-digit numbers, the answer is likely to be

in the ______________________

or ______________________

4. How large will the answer be if two, three-digit numbers are added?

tens	hundreds	thousands	bigger

5. Create some three-digit numbers and add them.

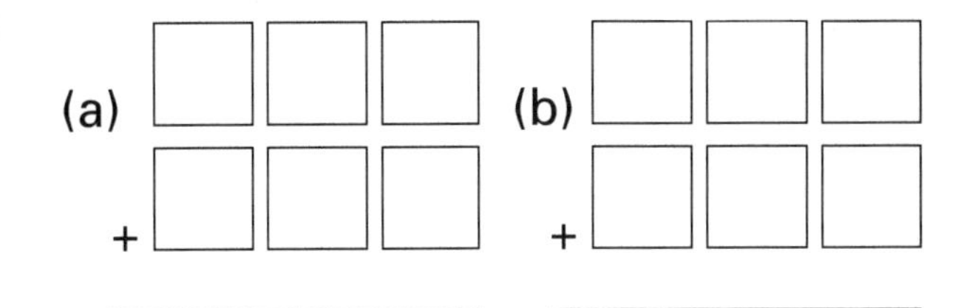

6. Lewis completed the following addition questions and made a few mistakes. Without completing the calculation again, circle the questions that you think he got wrong.

(a) 38 + 41 = 79 (c) 109 + 211 = 320 (e) 970 + 620 = 159

(b) 86 + 47 = 1,213 (d) 132 + 47 = 85 (f) 840 + 380 = 1,220

Explain your answers.

____

____

____

7. On the back of this sheet:

Investigate what happens if you add two four-digit numbers. What patterns do you notice?

or

Investigate what happens when you subtract numbers. Start by trying combinations of two-digit subtractions. Write about your findings.

DIVISIBILITY RULES

The focus of this section is on observing and exploring patterns that lead to the development of the divisibility rules. Some of the divisibility rules are simpler than others, so it is not expected that all students learn all the rules. The real value of these activities lies in the discovery of the patterns that lead to the development of the divisibility rules. The application of these rules is simply the end product.

If students are just told to learn the divisibility rules and apply them to checking their work, much of the value of these activities will be lost.

Divisibility Rules

It is often helpful to know when one number may be divided into another neatly without leaving a remainder.

Two

Any even number (a number that ends in 2, 4, 6, 8, or 0) will divide by two. Any number multiplied by two should end in 2, 4, 6, 8, or 0. This means that any number, no matter how large, multiplied by any number ending in two, should produce a result that ends in 2, 4, 6, 8, or 0.

That means if I multiply 7,369,112 by 6,337 and the result is not even, I have made a mistake. In fact the answer should end in a 4 because 2 x 7 = 14.

Three

A pattern in the multiples of three can be used to determine whether a number is divisible by three without leaving any remainders. If you add all the digits in the number and the total divides by three, the number will divide by three.

For example, 76,511 is not divisible by three because 7 + 6 + 5 + 1 + 1 = 20 which is not divisible by three without leaving a remainder.

8,652 is divisible by three because 8 + 6 + 5 + 2 = 21 and 21 is divisible by three.

In fact, 21 → 2 + 1 = 3

That means any number multiplied by a multiple of three (i.e. 3, 6, 9, 12, 15 …) should produce a result that is divisible by three without leaving a remainder.

Four

A number is divisible by 4 if the last two digits form a number that is divisible by 4.

For example, 23,632 is divisible by 4, without leaving a remainder, because 32 is divisible by 4.

4, 8, 12, 16, 20, 24, 28, 32, 36, 40, 44, 48, 52, 56, 60, 64, 68, 72, 76, 80, 84, 88, 92, 96, 00

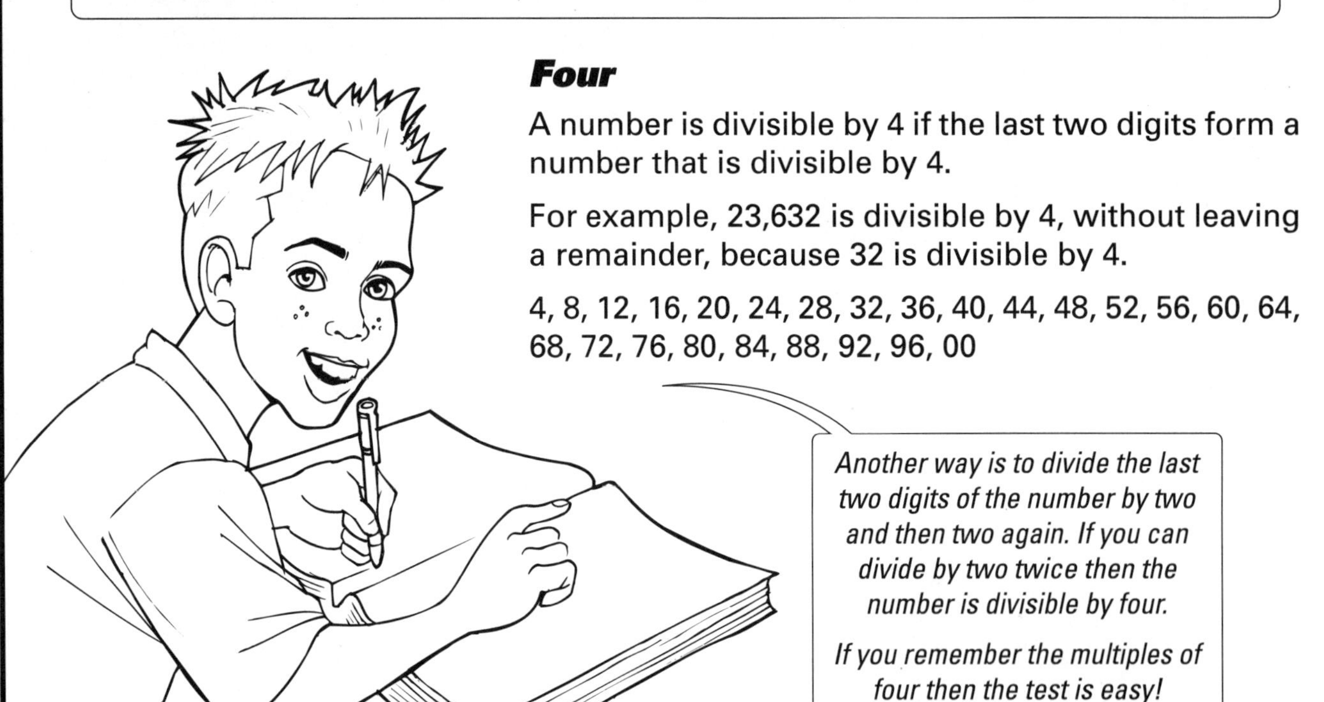

Five

This rule is easy and relies on observing the 0, 5, 0, 5 pattern when numbers are multiplied by 5 or a multiple of 5. Any number that ends in a zero or five will be divisible by five without leaving any remainders.

This means any number multiplied by 5, or a multiple of 5 will end in a 5 or 0. Therefore, 7,632,116 x 725 has to end in 5 or 0.

In fact, it should end in 0 because 6 x 5 = 30.

Six

The rule for divisibility by six is a little more tricky and relies on knowing the divisibility checks for two and three (2 x 3 = 6). If the number meets both the criteria for divisibility by two and three it will be divisible by six.

Seven

The divisibility check for seven is quite tricky and probably too complex for most students.

Eight

A number is divisible by 8 if the number formed by the last three digits is divisible by 8. Once again this check would be beyond the ability of most students.

Nine

The check for divisibility by nine is similar to the check for divisibility by three and quite easy to perform. Add all the digits in the number to see whether nine or a multiple of nine is formed. If nine or a multiple is formed, the number is divisible by nine.

For example, 72,567 → 7 + 2 + 5 + 6 + 7 = 27, and 2 + 7 → 9.

So the number is divisible by nine without any remainders.

Note: 27 → 2 + 7 = 93

That means any number multiplied by nine or a multiple of nine will produce a result that is divisible by nine.

Ten

Ten is easy. The number has to end in zero!

Eleven

Alternative digit sums differ by a multiple of 11 or 0.

For example, 76,285

$$\begin{aligned} 7 + 2 + 5 &= 14 \\ 6 + 8 &= 14 \\ 14 - 14 &= 0 \end{aligned}$$

Therefore, this number is divisible by 11.

Twelve

Both the rules for three and four have to apply for the number to be divisible by twelve.

Twenty-five

If the last two digits are divisible by 25; for example, 6,475.

Hundred

The last two digits of the number are both 0. For example, 67,400.

Cut along dotted line and join to page 22.

Divisibility rules

Calculator counting (+2)

Teacher information

Calculators are the ideal tool for generating data rapidly, which allows students to focus on the patterns that are produced. The constant function found on most calculators may be used to generate a variety of patterns. In this case, the calculator is set to count by two:

+ → 2 → = → = → = …

Note: Different calculator models may require an alternate set of keystrokes such as:

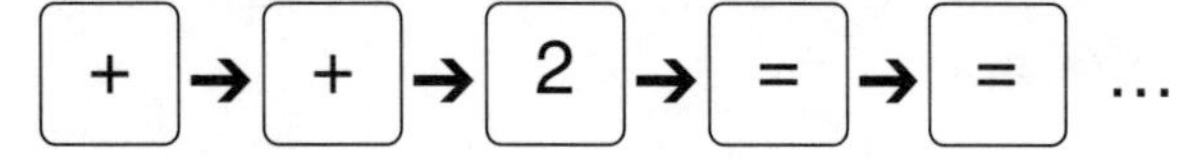

You will need to experiment with your calculator prior to trying this activity. Setting the calculator to count by two will focus students' attention on the even numbers. Some students will need to be prompted to notice the pattern, hence the suggestion of placing a finger over most of the display. This is further reinforced by shading the number grid.

Answers

1. 16
2. Multiply two by eight
3. Multiply 2 by 20
4. Multiply 2 by 76
5. 2, 4, 6, 8, 0, 2, 4 …
6. (a) All the even numbers are shaded. Every second column is shaded.
(b) A number is divisible by 2 if it is an even number; i.e. ends in 0, 2, 4, 6 or 8.
(c) If a number has been doubled the last digit will be 0, 2, 4, 6 or 8.

Calculator counting (+ 2)

To enable your calculator to count by two, try pressing [+]→[2]→[=]→[=] … The display should show 2, 4, 6, 8 … . If not, you may need to try a different sequence. When you press "=" it is like adding another two to the number shown in the display. If you press [+]→[2]→[=]→[=]→[=]→[=]→[=]→[=], you will end up with 12 showing on the display. Pressing [=] six times is like adding 2 six times.

1. What number will be shown on the display if you begin with 2 and press [=] eight times?

   ____

2. If you wanted to know the answer to eight lots of two, instead of adding 2 eight times to show 16 on the display, what else could you do?

   ____

3. Describe a quick way of adding 2 twenty times.

   ____

4. Instead of adding 2 seventy-six times, what could you do?

   ____

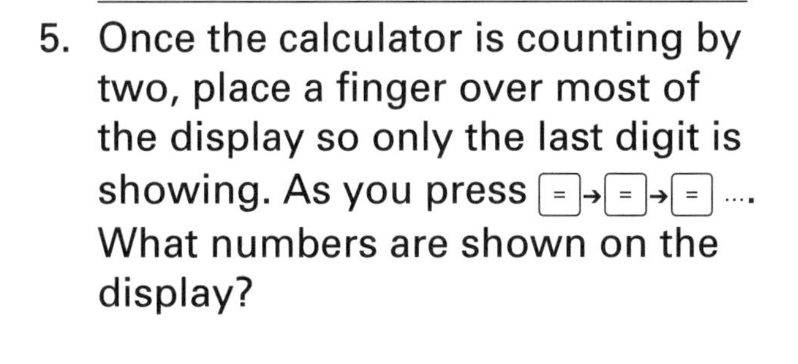

5. Once the calculator is counting by two, place a finger over most of the display so only the last digit is showing. As you press [=]→[=]→[=] …. What numbers are shown on the display?

   ____

   ____

6. Set your calculator counting by 2 and shade in the 0 – 99 grid each time you press [=].

(a) What do you notice?

____

____

(b) How could you tell if a number is divisible by two?

____

____

(c) If a number has been doubled, what might be the last digit?

____

____

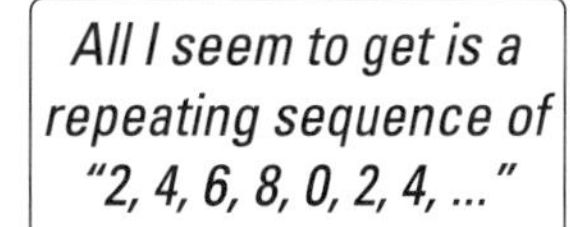

0	1	2	3	4	5	6	7	8	9
10	11	12	13	14	15	16	17	18	19
20	21	22	23	24	25	26	27	28	29
30	31	32	33	34	35	36	37	38	39
40	41	42	43	44	45	46	47	48	49
50	51	52	53	54	55	56	57	58	59
60	61	62	63	64	65	66	67	68	69
70	71	72	73	74	75	76	77	78	79
80	81	82	83	84	85	86	87	88	89
90	91	92	93	94	95	96	97	98	99

Divisibility rules

Calculator counting (+5, +10)

Teacher information

Setting the calculator to count by 5 will help to focus students on the 5, 0, 5, 0 pattern that is formed. Likewise, when counting by 10 students should notice that the units digit is always "0."

The constant feature of the calculator may be used to examine further patterns. For example, by pressing 2 + 5 = = =, the calculator display will show 7, 12, 17, 22, 27, 32.

Students could be asked to describe the pattern. Similarly, by pressing 97 – 5 = = =, students could observe the pattern 92, 87, 82 and 77 generated on the display of the calculator.

Answers

1. *40*
2. *Multiply 5 by 8*
3. *Multiply 5 by 20*
4. *137 x 5*
5. *(a) 13 x 4*
   (b) 11 x 17
   (c) Multiplication is repeated addition.
6. *They all end in 5 or 0.*
7. *5 or 0*
8. *It must end in 5 or 0.*
9. *The last digit is always 0.*
10. *0*
11. *It must end in 0.*
12. *Numbers divisible by 5 must end in 5 or 0; Numbers divisible by 10 must end in 0. When a number is multiplied by 5 it should end in 5 or 0.*

Calculator counting (+5, +10)

To enable your calculator to count by five, try pressing [+]→[5]→[=]→[=] .... The display should show 10. If not, you may need to try a different sequence. When you press [=] it is like adding another five to the number shown on the display. If you press [+]→[5]→[=]→[=]→[=]→[=]→[=]→[=], you should have 30 showing on the display. Pressing [=] six times is like adding 5 six times.

1. What number do you think would be shown on the display if you begin with five and press [=] eight times?

 ____

2. If you wanted to know the answer to eight lots of five, instead of adding 5 eight times to show 40 on the display, what else could you do?

 ____

3. Describe a quick way of adding 5 twenty times.

 ____

4. Describe a quick way of adding 5 one hundred thirty-seven times.

 ____

5. Explain a quick way to complete the following calculations.

 (a) 4 + 4 + 4 + 4 + 4 + 4 + 4 + 4 + 4 + 4 + 4 + 4 + 4

 ____

 (b) 17 + 17 + 17 + 17 + 17 + 17 + 17 + 17 + 17 + 17 + 17

 ____

 (c) Explain how addition is related to multiplication.

 ____

 ____

 ____

 ____

 ____

6. Once the calculator is counting by five, place a finger over most of the display so that only the last digit is showing. As you press [=]→[=]→[=] ..., note what numbers are shown on the display.

 ____

 ____

7. If a number has been multiplied by five, what might be the last digit?

 ____

8. How could you tell if a number is divisible by five?

 ____

Set your calculator to count by 10s. On most calculators you need to press [+]→[1][0]→[=]→[=]→[=]→[=]. Place a finger over most of the display so that only the last digit is showing.

9. What do you notice about the last digit in the display?

 ____

10. If a number has been multiplied by 10, what will be the last digit?

 ____

11. How can you tell if a number is divisible by ten?

 ____

12. On the back of this sheet, explain how what you have learned about multiplying and dividing by 5 and 10 might help you check your work.

Divisibility rules

Dividing by three – 1

Teacher information

Every number has a digit sum. The digit sum is found by adding all the digits that make up a number, until a single digit is left. For example, the digit sum of 7 is seven; The digit sum of 62 is eight (6 + 2); The digit sum of 728 is also eight (7 + 2 + 8 = 17, 1 + 7 = 8).

The numbers that are divisible by three, without leaving a remainder, exhibit an interesting characteristic—the digit sum is always 3, 6, or 9.

Answers

1. (a) *yes*
   (b) *no*
   (c) *no*
   (d) *yes*
   (e) *yes*
   (f) *no*
   (g) *no*
   (h) *yes*
   (i) *yes*
   (j) *no*
2. (d) *22 → 2 + 2 → 4*
   (e) *4 + 5 + 6 + 7*
   (f) *18 → 1 + 8 → 9*
   (g) *9 + 5 + 2 + 2*
   (h) *1 + 2 + 1 + 3 + 4 → 11 → 1 + 1 →2*
   (i) *1 + 5 + 0 + 0 + 0 → 6*
   (j) *2 + 1 + 0 + 0 + 3 → 6*
3. *The digit sums add up to 3, 6, or 9.*
4. *Teacher check.*
   A number will be divisible by three if the digit sum is 3, 6, or 9.

Dividing by three – 1

How can you work out whether a number divides by three without leaving a remainder?

1. Divide all the following numbers by three and note whether there are any remainders. Your calculator will show remainders after the decimal point.

For example,

460 ÷ 3 = 153.33333
461 ÷ 3 = 153.66666
462 ÷ 3 = 154
All the remainders will show up as either 0.333333 (which is $\dot{3}$) or 0.666666 (which is $\dot{6}$).

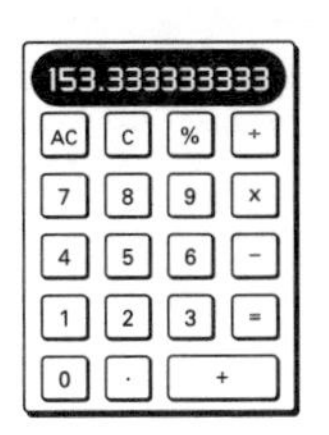

The little dot over the three and six means that it goes on forever!

(a) 731 ÷ 3 remainder → ◯ yes ◯ no
(b) 696 ÷ 3 remainder → ◯ yes ◯ no
(c) 3,456 ÷ 3 remainder → ◯ yes ◯ no
(d) 2,785 ÷ 3 remainder → ◯ yes ◯ no
(e) 4,567 ÷ 3 remainder → ◯ yes ◯ no
(f) 7,272 ÷ 3 remainder → ◯ yes ◯ no
(g) 9,522 ÷ 3 remainder → ◯ yes ◯ no
(h) 12,134 ÷ 3 remainder → ◯ yes ◯ no
(i) 15,002 ÷ 3 remainder → ◯ yes ◯ no
(j) 21,003 ÷ 3 remainder → ◯ yes ◯ no

2. Complete the missing parts of these number sentences.

7 + 3 + 1 → 11 → 1 + 1 → 2
6 + 9 + 6 → 21 → 2 + 1 → 3
3 + 4 + 5 + 6 → 18 → 1 + 8 → 9
2 + 7 + 8 + 5 → ☐ → ☐ + ☐ → ☐
☐ + ☐ + ☐ + ☐ → 22 → 2 + 2 → 4
7 + 2 + 7 + 2 → ☐ → ☐ + ☐ → ☐
☐ + ☐ + ☐ + ☐ → 18 → 1 + 8 → 9
☐ + ☐ + ☐ + ☐ + ☐ → ☐ → ☐ + ☐ → ☐
☐ + ☐ + ☐ + ☐ + ☐ → ☐
☐ + ☐ + ☐ + ☐ + ☐ → ☐

3. What do you notice about the numbers that are divisible by three and their digit sums?

____

4. Explain how what you have learned could help you check your work.

____
____
____
____

Calculate the digit sum.

Divisibility rules

Dividing by three – 2

Teacher information

Every number has a digit sum. The digit sum is found by adding all the digits that make up a number, until a single digit is left. For example, the digit sum of 7 is seven; The digit sum of 62 is eight (6 + 2); The digit sum of 728 is also eight (7 + 2 + 8 = 17, 1 + 7 = 8).

The numbers that are divisible by three, without leaving a remainder, exhibit an interesting characteristic—the digit sum is always 3, 6, or 9.

Students may note a second pattern. The digit sums of the multiples of three follow this pattern.

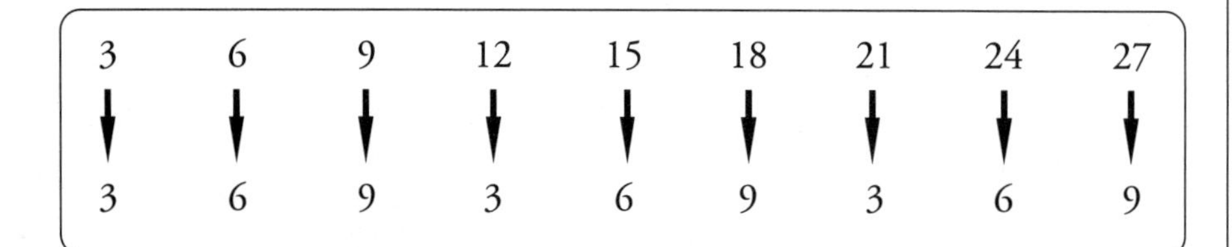

Answers

1. *3, 6, 9, 12, 15, 18, 21, 24, 27, 30, 33, 36, 39, 42, 45, 48, 51, 54, 57, 60, 63, 66, 69, 72, 75, 78, 81, 84, 87, 90, 93, 96, 99*
2. *They divide without leaving a remainder.*
3. (a) *8.66666 … ➔* $8.\dot{6}$
   (b) *12.3333 … ➔* $12.\dot{3}$
   (c) *13.6666 … ➔* $13.\dot{6}$
   (d) *18.3333 … ➔* $18.\dot{3}$
   (e) *20.6666 … ➔* $20.\dot{6}$
   (f) *24.3333 … ➔* $24.\dot{3}$
   (g) *29.3333 … ➔* $29.\dot{3}$
   (h) *33.3333 … ➔* $33.\dot{3}$
   (i) *There was a remainder in each.*
4. (a) *Teacher check*
   (b) *The pattern 3, 6, 9 continues. All numbers divisible by three reduce to a digit sum of 3, 6, or 9.*
   (c) *The digit sums for the numbers in question 3 are 8, 1, 5, 1, 8, 1, 7 and 1*
   (d) *None of the digit sums came to 3, 6, or 9.*
5. *By adding the digits to see whether the result is 3, 6, or 9.*
6. (a) *no* $1 + 3 + 4 = 8$
   (b) *yes* $1 + 7 + 7 = 15, 1 + 5 = 6$
   (c) *yes* $2 + 0 + 1 = 3$
   (d) *no* $2 + 3 + 3 = 8$
   (e) *no* $3 + 2 + 2 = 7$
   (f) *yes* $3 + 5 + 1 = 9$
7. *The digits 2, 3 and 7 add to 12, 1 + 2 = 3, so as long as the two missing digits also add to 3, 6, or 9, then the entire five-digit will be divisible by three without leaving a remainder.*

   Some possible digits include:

   3 ➔ 1, 2 and 21, 23,127 or 23,217
   3 ➔ 3, 0 and 0, 3, 23,307 or 23,037
   6 ➔ 1, 5 and 5, 1, 23,157 or 23,517
   6 ➔ 3, 3 23,337
   9 ➔ 1, 8 and 8, 1, 23,187 or 23,817
   9 ➔ 2, 7 and 7, 2, 23,277 or 23,727
   9 ➔ 3, 6 and 6, 3, 23,367 or 23,637
   9 ➔ 4, 5 and 5, 4, 23,457 or 23,547

Dividing by three – 2

1. Write all the multiples of three to 100.

____

____

____

____

____

____

____

2. Divide all these numbers by three. What do you notice?

____

____

____

3. Divide the following numbers by three.

   (a) 26 ____

   (b) 37 ____

   (c) 41 ____

   (d) 55 ____

   (e) 62 ____

   (f) 73 ____

   (g) 88 ____

   (h) 100 ____

   (i) What do you notice?

   ____

   ____

   ____

4. *(a) On the back of this sheet add the digits of each of your original multiples of three until a single digit number is formed.*

   For example, 48 → 4 + 8 → 12 → 1 + 2 → 3

   (b) What do you notice? ______________________

   ____

   ____

   (c) Now try the numbers in question 3.

   (d) What do you notice? ______________________

   ____

5. How could you tell if a number is divisible by three just by looking at it?

____

____

6. Which of the following numbers do you think are divisible by three without leaving a remainder? Test your theory.

   (a) 134 ◯ *yes* ◯ *no* ______________________

   (b) 177 ◯ *yes* ◯ *no* ______________________

   (c) 201 ◯ *yes* ◯ *no* ______________________

   (d) 233 ◯ *yes* ◯ *no* ______________________

   (e) 322 ◯ *yes* ◯ *no* ______________________

   (f) 351 ◯ *yes* ◯ *no* ______________________

7. Two digits are missing from this five-digit number.

   2 3, ___ ___ 7

   The number is divisible by three without leaving a remainder. List some possible numbers that fit this description. Explain how you arrived at your answer.

____

____

____

Divisibility rules

Noticing nines

Teacher information

The digit sum of multiples of nine is always nine. Once students are familiar with this pattern, investigations such as the following one may be set.

The following five-digit number was divisible by nine without leaving a remainder. Unfortunately, two of the digits have been erased.

71,__6__

What number might have been written down?

7 + 1 + 6 → 14 → 1 + 4 → 5

The missing digits must therefore add to four or thirteen for the numbers to be divisible by 9.

The digits must add to nine but currently only add to five.

Answers

71,163	71,361	71,262
71,460	71,064	71,469
71,964	71,568	71,667
71,766		

Answers

1. *Digit sum for all is nine.*
2. *Numbers divisible by nine have a digit sum of nine. Also numbers multiplied by nine will have a digit sum of nine.*
3. (a) *no; 8 + 3 + 1 → 12 → 1 + 2 → 3*
   (b) *no; 6 + 9 + 6 → 21 → 2 + 1 → 3*
   (c) *yes; 3 + 4 + 5 + 6 → 18 → 1 + 8 → 9*
   (d) *no; 2 + 7 + 8 + 5 → 22 → 2 + 2 → 4*
   (e) *no; 4 + 5 + 6 + 7 → 22 → 2 + 2 → 4*
   (f) *yes; 7 + 2 + 7 + 2 → 18 → 1 + 8 → 9*
   (g) *yes; 9 + 5 + 2 + 2 → 18 → 1 + 8 → 9*
   (h) *no; 1 + 2 + 1 + 3 + 4 → 11 → 1 + 1 → 2*
   (i) *no; 1 + 5 + 0 + 0 + 2 → 8*

Noticing nines

1. Write down the nine times table.

(a) 1 x 9 = ____

(b) 2 x 9 = ____ → 8 + 1 = ☐

(c) 3 x 9 = ____ → ____ = ☐

(d) 4 x 9 = ____ → ____ = ☐

(e) 5 x 9 = ____ → ____ = ☐

(f) 6 x 9 = ____ → ____ = ☐

(g) 7 x 9 = ____ → ____ = ☐

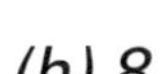

(h) 8 x 9 = ____ → ____ = ☐

(i) 9 x 9 = ____ → ____ = ☐

(j) 10 x 9 = ____ → ____ = ☐

(k) 11 x 9 = ____ → ____ = ☐

(l) 12 x 9 = ____ → ____ = ☐

(m) 13 x 9 = ____ → ____ = ☐

(n) 14 x 9 = ____ → ____ = ☐

(o) 15 x 9 = ____ → ____ = ☐

(p) 16 x 9 = ____ → ____ = ☐

(q) 17 x 9 = ____ → ____ = ☐

(r) 18 x 9 = ____ → ____ = ☐

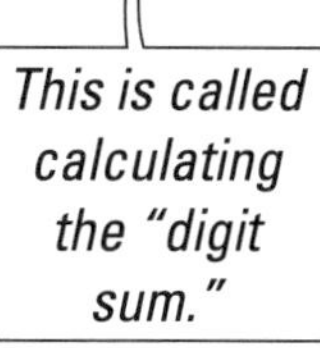

2. How can knowing about this pattern help you check your work?

____

____

3. Decide which of the following numbers are divisible by nine without leaving a remainder and then test your predictions.

(a) 831 ◯ yes ◯ no ____________ ◯ correct?

(b) 696 ◯ yes ◯ no ____________ ◯ correct?

(c) 3,456 ◯ yes ◯ no ____________ ◯ correct?

(d) 2,785 ◯ yes ◯ no ____________ ◯ correct?

(e) 4,567 ◯ yes ◯ no ____________ ◯ correct?

(f) 7,272 ◯ yes ◯ no ____________ ◯ correct?

(g) 9,522 ◯ yes ◯ no ____________ ◯ correct?

(h) 12,134 ◯ yes ◯ no ____________ ◯ correct?

(i) 15,002 ◯ yes ◯ no ____________ ◯ correct?

Divisibility rules

Division discovery

Teacher information

This activity reinforces the rules for divisibility by three and nine. In addition, students should note that all the numbers divisible by nine are also divisible by three, but not all the numbers divisible by three are also divisible by nine, without leaving a remainder.

After completing this activity you may like to set the following extension to challenge the students.

1. *Choose a two-digit number that is not divisible by three and enter it into your calculator. (If you are not sure whether the number you have chosen is divisible by three, check with your calculator.)*
2. *Repeat the two-digit character so that a six-digit number is formed.*
3. *Divide the six-digit number by three.*
4. *Write about what you notice.*
5. *Try again with some other two-digit numbers that are not divisible by three.*
6. *Try to explain why this happens.*
7. *Is there a shortcut method of finding out whether a number is divisible by three?*

Answer notes

This is based on the rule of divisibility by three, which basically states that a number is divisible by three if the sum of its digits is divisible by 3. Repeating a two-digit number twice means that you will end up with three lots of the original two-digit number or three times the digital sum; therefore the six-digit number must be divisible by three.

Answers

1.

Number	*Divisible by 3*	*Divisible by 9*	*Digit sum*	*Digit sum reduces to 3 or a multiple of 3*	*Digit sum reduces to 9*
84	yes	no	8+4 → 12 → 1+2 → 3	yes	no
117	yes	no	1+1+7 → 9	yes	yes
222	yes	no	2+2+2 → 6	yes	no
235	no	no	2+3+5 → 10 → 1+0 → 1	no	no
333	yes	no	3+3+3 → 9	yes	yes
372	yes	no	3+7+2 → 12 → 1+2 → 3	yes	no
432	yes	no	4+3+2 → 9	yes	yes
510	yes	no	8+4 → 12 → 1+2 → 3	yes	no
741	yes	no	7+4+1 → 12 → 1+2 → 3	yes	no
744	yes	no	7+4+4 → 15 → 1+5 → 6	yes	no
747	yes	no	7+4+7 → 18 → 1+8 → 9	yes	yes
23,454	yes	no	2+3+4+5+4 → 18 → 1+8 → 9	yes	yes

2. *No*
3. *Yes*
4. *Teacher check*
5. *A digit sum of three, six or nine indicates the number will be divisible by three without leaving a remainder.*
   A digit sum of nine indicates the number will be divisible by nine without leaving a remainder.

Q. What did you notice about the numbers 741, 744, 747?
A. *They are three apart and all divisible by three.*

Division discovery

Some of the numbers in the following table are divisible by three without leaving a remainder and some are divisible by nine.

1. Complete the table and note which numbers are divisible by three and which are divisible by nine.

Number	Divisible by 3	Divisible by 9	Digit sum	Digit sum reduces to 3 or a multiple of 3	Digit sum reduces to 9
84	yes	no	8 + 4 → 12 1 + 2 → 3	yes	no
117					
222					
235					
333					
372					
432					
510					
741					
744					
747					
23,454					

2. If a number is divisible by three without leaving a remainder, will it also be divisible by nine without leaving a remainder?

What did you notice about the numbers 741, 744 and 747?

3. If a number is divisible by nine without leaving a remainder, will it also be divisible by three without leaving a remainder?

4. Color all the multiples of nine on the number chart. Use a different color to mark all the multiples of three. What do you notice?

5. On the back of this sheet explain how the digit sum helps you to work out whether the number will be divisible by 3 or 9 without leaving a remainder.

0	1	2	3	4	5	6	7	8	9
10	11	12	13	14	15	16	17	18	19
20	21	22	23	24	25	26	27	28	29
30	31	32	33	34	35	36	37	38	39
40	41	42	43	44	45	46	47	48	49
50	51	52	53	54	55	56	57	58	59
60	61	62	63	64	65	66	67	68	69
70	71	72	73	74	75	76	77	78	79
80	81	82	83	84	85	86	87	88	89
90	91	92	93	94	95	96	97	98	99

Divisibility rules

Dividing by four – 1

Teacher information

Marking the multiples of four and listing them will raise awareness of them and hence assist students to recognize them. This in turn will help students when examining numbers to determine whether they will be divisible by four, without leaving a remainder.

Students should be encouraged to consider what happens when a number is multiplied by four or a multiple of four.

For example, 57 x 24 = 1,3<u>68</u>

The last two digits are a multiple of four.

Answers

1.

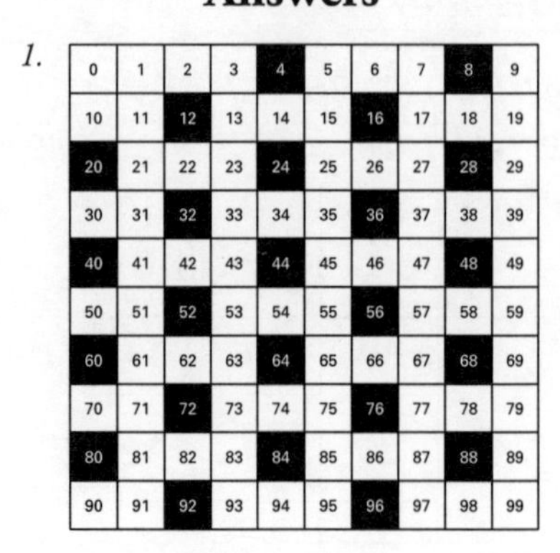

0	1	2	3	**4**	5	6	7	**8**	9
10	11	**12**	13	14	15	**16**	17	18	19
20	21	22	23	**24**	25	26	27	**28**	29
30	31	**32**	33	34	35	**36**	37	38	39
40	41	42	43	**44**	45	46	47	**48**	49
50	51	**52**	53	54	55	**56**	57	58	59
60	61	62	63	**64**	65	66	67	**68**	69
70	71	**72**	73	74	75	**76**	77	78	79
80	81	82	83	**84**	85	86	87	**88**	89
90	91	**92**	93	94	95	**96**	97	98	99

2. *4, 8, 12, 16, 20, 24, 28, 32, 36, 40, 44, 48, 52, 56, 60, 64, 68, 72, 76, 80, 84, 88, 92, 96*
3. *Answers will vary*
4. *A number is divisible by four if the last two digits are divisible by 4.*

Dividing by four – 1

1. Mark all the multiples of four on the number chart below.

0	1	2	3	4	5	6	7	8	9
10	11	12	13	14	15	16	17	18	19
20	21	22	23	24	25	26	27	28	29
30	31	32	33	34	35	36	37	38	39
40	41	42	43	44	45	46	47	48	49
50	51	52	53	54	55	56	57	58	59
60	61	62	63	64	65	66	67	68	69
70	71	72	73	74	75	76	77	78	79
80	81	82	83	84	85	86	87	88	89
90	91	92	93	94	95	96	97	98	99

2. List all the multiples of four.

____

____

____

____

____

____

____

____

3. Write some three-digit numbers and then add any of the two-digit multiples of four to the end of the three-digit number to produce a five-digit number. Divide this five-digit number by 4. What do you notice?

three-digit number	*two-digit multiple of 4*	*combined five-digit number*	*divisible by 4*
231	24	23,124	yes

4. How can you tell whether a number is divisible by four without using a calculator? ____

____

Divisibility rules

Dividing by four – 2

Teacher information

A number is divisible by four if the last two digits form a number that is divisible by four. For example, 36,764 will be divisible by four without leaving a remainder because 64 is divisible by four.

Students are not expected to memorize the rules for divisibility, especially those associated with four, but discussion about the patterns associated with four should heighten awareness.

Students should recognize that a number multiplied by four will end in a multiple of four.

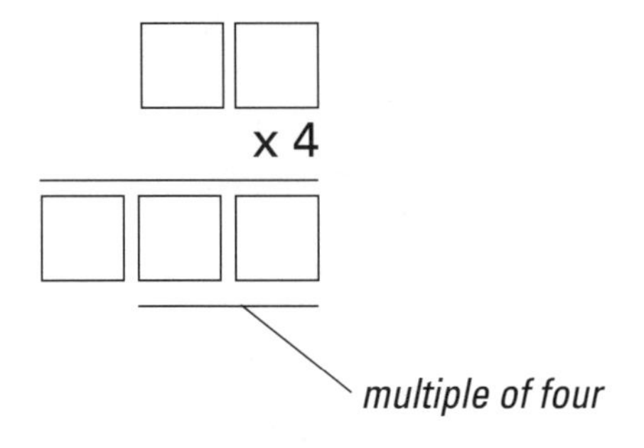

Or ________ x a multiple of four will produce a number divisible by four.

Answers

1.

16	yes	16
32	yes	32
82	no	82
116	yes	16
122	no	22
136	yes	36
254	no	54
288	yes	88
312	yes	12
316	yes	16
318	no	18
535	no	35
648	yes	48
676	yes	76
682	no	82
1,020	yes	20
1,042	no	42

2. *They are all multiples of four.*
3. *Teacher check*

Dividing by four – 2

1. Complete the following table and note which numbers are divisible by four without leaving a remainder.

Number	÷ 4	Is it divisble by four without leaving a remainder?	Last two digits of original number.
16	4	yes	16
32	8	yes	32
82			
116			
122			
136			
254			
288			
312			
316			
318			
535			
648			
676			
682			
1,020			
1,042			

2. What do notice about the last two digits of the numbers that are divisible by four (without leaving a remainder)?

____

____

____

____

____

3. Make up some four-digit numbers, where the last two digits are multiples of four—e.g. 2,1<u>52</u>, 2,2<u>76</u>, 3,7<u>92</u>—and divide them by four. What do you notice?

____

____

____

____

Divisibility rules

Dividing by eight

Teacher information

The rule for divisibility by eight is quite difficult for students to discover; hence they are given the rule and asked to test it. Students are not expected to commit this rule to memory. It is included as an extension of the work on divisibility by four.

Class discussion should lead to the discovery that numbers divisible by eight without leaving a remainder will also be divisible by two and four. A number that is divisible by four will also be divisible by two. A number divisible by four, without leaving a remainder, however, is not necessarily divisible by eight without leaving a remainder. Likewise, a number divisible by two is not necessarily divisible by eight. Examining these patterns heightens student number intuition.

Read the examples below to students or groups who need guidance with Question 1.

I am going to press + 8 = = = = on my calculator and when I get past 100 I am going to list some of the three-digit numbers that come up in the display. Then I will use these numbers as the last three digits of some larger numbers that I make up. Then I can test whether these numbers are divisible by eight without leaving a remainder.

I am just going to multiply some numbers by eight to make some three-digit numbers. These numbers will have to be divisible by eight. I will then make up some bigger numbers using the three-digit numbers as the last three digits. I can then check whether the larger numbers are divisible by eight without leaving a remainder.

Answers

1. *Teacher check*

2 – 4. *Teacher check*
A number is divisible by eight only if the last three digits of the number are divisible by eight.

Dividing by eight

*A forgetful mathematics professor once told me that you could tell whether a number was divisible by eight without leaving a remainder by looking at the last three digits of the number. She said that a number is divisible by eight if and **only if** the last three digits of the number are divisible by eight.*

1. Design an experiment that will help you to check whether this statement is true.
2. Describe what you have found. ______________________________

____

____

I also heard there are some other patterns that might help you to find out whether the last three digits are divisible by eight. For example, I have heard that you can tell if the last three digits are divisible by eight by looking at the first left digit (the number in the hundreds place) to see whether it is even. If the first digit is even then you only need to look at the last two digits to see if they are divisible by eight.

3. Write examples to test this theory.

I wonder if this works!

Let me see ... 72 is divisible by 8 because 9 x 8 = 72, but 172 is not evenly divisible by eight. 272 is—372 isn't!

I will have to try a few more examples to test this theory!

4. Explain your findings. ______________________________

____

____

____

ESTIMATION AND ROUNDING

The term estimation embraces a variety of methods used to calculate a rough answer to a computation so that a comparison can be made for checking purposes. Labato (1993) recognizes the complex nature of the act of estimating when defining it as *guessing with a little bit of problem solving.*

Labato, J.E. (1993) Making connections with estimation. Arithmetic Teacher, 40 (6). 347–351

How to improve estimation

Estimation is a rather complex process that involves the use of problem-solving skills; hence the ability to estimate will vary from student to student. Good estimators, however, tend to possess certain characteristics which assist them to make an estimate. Having a good grasp of basic facts, place value and arithmetic properties, which in turn assists with mental computation, is fundamental to being able to estimate. Basic number fact knowledge and associated knowledge may be combined with flexible thinking (problem-solving ability) and the use of some simple strategies to form estimates.

It is not the purpose of this book to review basic fact knowledge, place value or properties of arithmetic, but to improve their ability to estimate, students need these skills and understandings. The focus of the next section will be on some of the simple strategies students use in combination with the above mentioned skills and understandings to assist in the formation of an estimate.

Estimation and rounding

Teacher information

There are several different techniques that may be used for estimating. Some techniques are more useful for particular operations. For example, when adding two numbers or two-digit numbers a simple rounding procedure may be all that is required, but this may become cumbersome when adding larger numbers or when adding more than two numbers. Remember, unless an estimate may be made quickly and without too much effort, students will not try.

Front-end rounding

A simple form of estimation for adding a series of numbers is called front-end rounding. There are two important steps when using this technique.

1. Look at the left-most digit in a number.
2. Consider the place-value of the digit 3.

❸215
❻910
+ ❹342

3 + 6 + 4 = 13

So my estimate would be 13,000.

For example, when adding you would simply ignore all the digits and then make use of place value to produce an estimate. Later on, students may be encouraged to consider the effect the 900 has on the estimate.

It should be noted that students find this technique fairly simple to use as it involves covering the digits to the right of the left-most digit. As the students become more proficient in the use of this method they should be encouraged to look at the digits in the next place along in order to produce an estimate closer to the actual result. This technique relies on students being able to add numbers beyond the basic facts. There are several ways the front-end strategy may be varied, including the clustering of numbers according to their magnitudes; i.e. numbers in the hundreds, thousands, etc.

(Note: The front-end rounding technique will always produce an underestimate.)

Students need to be taught a couple of basic estimation techniques such as front-end rounding and rounding and then be given the opportunity to experiment with these methods to determine which ones provide closer estimates in various circumstances. The size of the numbers involved, the level of rounding and the operation involved will all play a role in selecting an estimation technique. Students who experience difficulty in estimating will often choose to use the front-end methods. While they may be able to produce closer estimates using rounding methods, at least students who apply front-end methods are thinking about the numbers involved and the possible answer.

A brief summary of these two main approaches to estimation is given below.

Method	Front-end rounding	Rounding
Addition	Always leads to an underestimate; simple and quick; may be adjusted to get closer to the answer.	Slower—may provide a closer estimate.
Subtraction	Often produces estimates well above or below the actual answer.	Tends to produce estimates closer to the result; generally, fairly easy to use.
Multiplication	Will produce estimates below the actual answers. At times, these will be significantly below the real answer.	Estimates can vary considerably from the actual answer; tends to give a closer estimate than front-end methods.
Division	Front-end methods tend to tie in well with division as it is performed from left to right unlike other operations. Will lead to an underestimate.	Rounding needs to take into account the relationships between the numbers. For example, when rounding 632 ÷ 8 a student might choose to round to 640 rather than 630 as 640 is easily divided by 8.

Estimation and rounding

Estimation

Teacher information

This first activity is designed to show that estimates generally fall within certain limits. If an estimate is too broad it may not be helpful. At times, however, all you may need is an estimate of "too much," "not enough," or "between … and …"

For example, when buying groceries you may deliberately overestimate the value of the goods to avoid running short of the cash to pay at the counter.

Answers

1. (a) *less than*
   (b) *more than*
   (c) *more than*
   (d) *Teacher check*
2. (a) *less than*
   (b) *more than*
   (c) *less than*
   (d) *Teacher check*
3. (a) *less than*
   (b) *more than*
   (c) *less than*
   (d) *Teacher check*
4. (a) *less than*
   (b) *more than*
   (c) *more than*
   (d) *Teacher check*
5. (a) *no*
   (b) *Teacher check*

Estimation

Decide whether the following calculations will be ***more*** or ***less*** than the number that is given.

Example: 61
+ 74
<u>more than</u> 100 or *less than* 100

1. (a) $57 + 94$ ______ 200 (b) $61 + 58$ ______ 100 (c) $117 + 98$ ______ 200

   (d) *Explain how you decided whether the answer was more than or less than the given number.*

   ____

2. (a) $656 - 490$ ______ 200 (b) $798 - 256$ ______ 500 (c) $916 - 582$ ______ 500

   (d) *Explain how you decided whether the answer was more than or less than the given number.*

   ____

3. (a) 25×18 ______ 500 (b) 27×34 ______ 900 (c) 41×38 ______ 1,600

   (d) *Explain how you decided whether the answer was more than or less than the given number.*

   ____

4. (a) $2,070 \div 52$ ______ 40 (b) $3,714 \div 89$ ______ 40 (c) $491 \div 16$ ______ 30

   (d) *Explain how you decided whether the answer was more than or less than the given number.*

   ____

5. (a) *The crowd at a football game was reported as being 24,000. Do you think exactly 24,000 people attended the game?*

   ◯ *yes* ◯ *no*

   (b) *How many people do you think might have attended the game?*

   ____

Estimation and rounding

Rounding

Teacher information

The most common form of estimation involves the use of rounding. Various rules have often been applied to *rounding* but these often are devoid of any context. An element of rounding is required whenever a scale is read on a measuring device and the context in which the measurement is taken provides some clues as to how to round.

For example, it may make sense to round 273 to 270 and 1,942 to 1,940 or perhaps to 2,000, depending on the context. It may not, however, make sense to round 2 to 0. The context can help determine how to round.

Answers

1. *15,000*
2. *$2,000*
3. *1,000 miles*
4. *20 minutes*
5. *60 mph*
6. *10 o'clock*
7. *50,000 miles*
8. *13 cm*
9. *Answers will vary*

Rounding

Sometimes it is easier to use numbers that have been rounded. For example, sports commentators often estimate the size of a crowd of spectators. They might say 32,000 people attended a game, but in reality it could have been 31,670 or perhaps 32,416 people.

1. If 15,072 people attended a football game, what rounded number might be reported? ____________
2. $2,057 was raised for charity. What rounded amount might be reported?

   ____
3. A trip of 984 mi. might be described as being about ____________ mi. long.
4. A period of 20 minutes and 35 seconds might be described as

   ____________ minutes.
5. 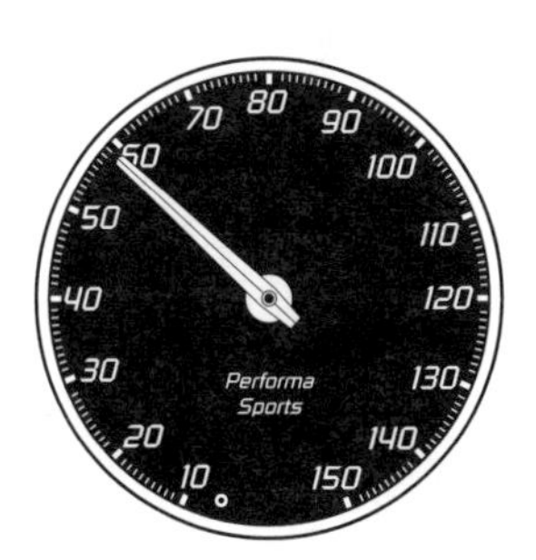

   The speed on the speedometer might be described as about

   ____________ mph.
6. 

   The time shown on the clock could be described as nearly

   ____________ o'clock.
7. The odometer on a used car reads

   The used car advertisement read "Good family vehicle. Traveled less than ____000 mi." What amount might be used?
8. 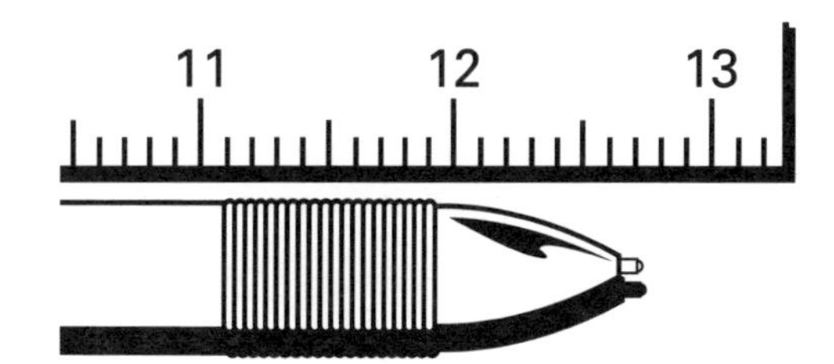

   When measuring a pen the length fell between 12 and 13 cm. This might be rounded to ____ cm.
9. 

   $100,000 Donated to Charity

   Story by Rory Masters

   Last night the late Lady Lydia of Lamington passed away leaving her fortune to her estranged son Marmaduke Lamington III. Mr Lamington had decided to donate $100,000 of his inheritance to charity to help people less fortunate than himself.

   What actual amount might have been donated to charity?

   ____

   ____

   ____

   ____

Estimation and rounding

Rounding to the nearest ...

Teacher information

Many real life situations involve rounding rather than the need for exact calculation. When shopping, most people overestimate to avoid the embarrassment of being caught short of cash (although with credit cards this is less of an issue). Rounding for the purposes of estimating in order to check the answer to a calculation is often devoid of a context and therefore presents a few issues.

The convention adopted in this book is for five to round to the next number; i.e. 45 rounded to the nearest 10 would be 50; 750 rounded to the nearest 100 would be 800, and so on.

Answers

1. (a) 40 (b) 30 (c) 50 (d) 40 (e) 50 (f) 60 (g) 70 (h) 70 (i) 80 (j) 80 (k) 90 (l) 90 (m) 100

2. (a) 110 (b) 120 (c) 120 (d) 130 (e) 130 (f) 210 (g) 310 (h) 390 (i) 400 (j) 400 (k) 650 (l) 540 (m) 1,000

3. (a) 400 (b) 400 (c) 700 (d) 700 (e) 700 (f) 800 (g) 800 (h) 800 (i) 800 (j) 800

4. Many numbers round to the same number, especially when rounding to the nearest 100.

5. Answers will vary but should be between 850 and 949.

6. (a) 200 (b) 900 (c) 1,000 (d) 3,500 (e) 7,000 (f) 7,000

Rounding to the nearest ...

When estimating to check an answer the numbers often need to be rounded first.

1. Round the following numbers to the nearest ten.

(a) 43 ________ (h) 72 ________

(b) 27 ________ (i) 77 ________

(c) 45 ________ (j) 81 ________

(d) 36 ________ (k) 88 ________

(e) 49 ________ (l) 93 ________

(f) 61 ________ (m) 95 ________

(g) 68 ________

43 has 4 tens. It could rounded to 4 tens or 5 tens. It is closer to 4 tens (40) than 5 tens (50) so it should be rounded to 40.

2. Round the following numbers to the nearest ten.

(a) 112 ________ (h) 389 ________

(b) 116 ________ (i) 399 ________

(c) 124 ________ (j) 401 ________

(d) 125 ________ (k) 650 ________

(e) 129 ________ (l) 536 ________

(f) 211 ________ (m) 997 ________

(g) 311 ________

112
Think how many tens the number has? (11)
What tens might it round to? (11 or 12)
Which ten is closer? (11)
112 rounds to 110

3. Round the following numbers to the nearest hundred.

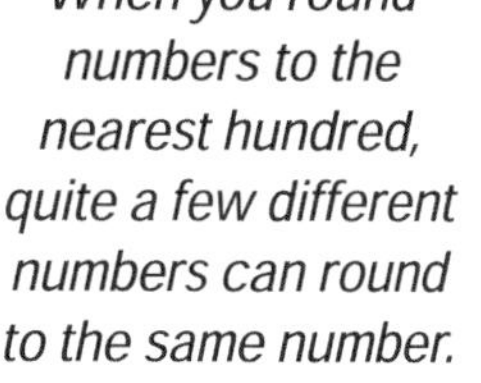

(a) 399 ________ (f) 751 ________

(b) 401 ________ (g) 775 ________

(c) 650 ________ (h) 789 ________

(d) 723 ________ (i) 799 ________

(e) 748 ________ (j) 802 ________

4. What do you notice about some of these numbers?

____

____

____

____

5. Write down some numbers that would round to 900.

____

____

____

____

6. Round these numbers to the nearest 100.

(a) 214 ________ (d) 3,516 ________

(b) 863 ________ (e) 7,043 ________

(c) 950 ________ (f) 6,975 ________

Estimation and rounding

Do I have enough?

Teacher information

Rather than insist on students using specific estimation techniques it would be more appropriate to present front-end methods and rounding methods and then allow them to experiment with each. After experimenting, students should be encouraged to share and discuss their methods. Remember, estimating may serve one of two purposes:

As a calculation method in its own right – when exact answers are not required
or
As a means of checking an exact calculation.

In this publication the focus is on the later and, therefore, when estimating students should be encouraged to use their estimates as a means of checking the result of an exact calculation. Thus, the techniques they adopt need to be quick and easy to execute.

Answers

1. *Yes ($8 – $9)*
2. *No (over $20)*
3. *Yes (around $40)*
4. *No (slightly over)*
5. *Over*
6. *No (over $11)*
7. *Yes (well under $20)*
8. *No (over $50)*
9. *Yes (slightly under)*
10. *Answers will vary*

Do I have enough?

When buying items at a supermarket it is a good idea to estimate whether you have enough money to pay for the items before reaching the checkout. Look at each cart, note the cost of each item and the amount of cash available, and determine whether there is enough money to cover the purchase.

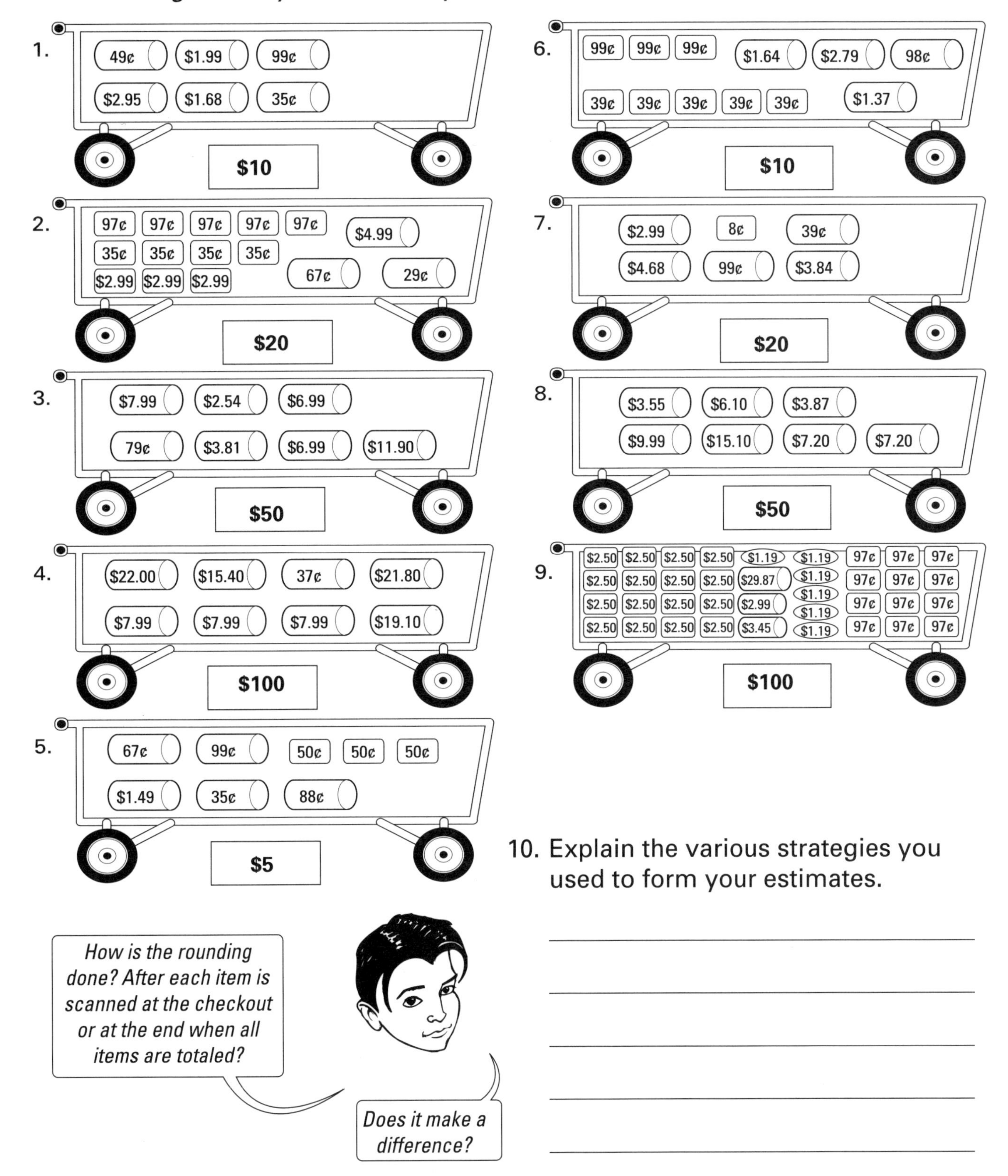

10. Explain the various strategies you used to form your estimates.

____

____

____

____

____

Estimation and rounding

That can't be right ... !

Teacher information

Context plays a role in estimating and checking work. The set activity involves the use of common sense and experience in real-world situations.

Students can work in pairs or small groups. Cut out each scenario and share them among the groups. Each group reads the scenario and discusses if it makes sense. Each scenario can be passed to another group after a few minutes. At the end of the lesson, discuss each scenario with the class. Ask groups to volunteer their responses.

Answers

1. *It does not make sense to ask for 4.376231 gal of paint when paint may only be bought in 1 and 5 gal cans. Absorption rates will vary, so to suggest an answer of 4.376231 gal is ridiculous.*
2. *The tank would have to be extremely small or the fuel very cheap. Twenty dollars would probably be more reasonable.*
3. *You can only pour 8, 250 mL glasses from a 2 L bottle. The answer is off by a factor of ten.*
4. *50 mL of medicine sounds like too high a dosage. Most medicine cups hold 5, 10, 15, or 20 mL. The dosage is probably out by a factor of 10 and should therefore be 5 mL.*
5. *Five buses is the correct result. Imagine how far a line of 50 buses would extend!*
6. *$847 for a cart-load of groceries sounds way too high. $84.70 or $184.70 might be more appropriate.*
7. *To appreciate what 30 m^3 of sand might look like, try building a one cubic meter model.*
8. *It is more likely that the girl runs the 100 m in 18.3 seconds. The current world record is around 10 seconds.*
9. *1 mile per minute is equal to 60 mph. 9 miles in 6 minutes equals 1.5 miles per minute or 90 mph. Not possible!*
10. *The average bag of potatoes weighs around 5 lb so it is more likely that the answer should be 84 lb.*
11. *5.7 inches is a little less than half a ruler length. The average height is much more likely to be 57 inches.*
12. *If a shopkeeper tried to charge $16.40 for just a few items I would check the receipt to see if any items had been incorrectly priced.*

That can't be right ... !

Have you ever looked at an answer to a mathematics problem and said, "That can't be right"? In real life there are certain clues that help us to determine when the answer may not be right. For example, most adults are between 140 cm and 190 cm tall, so if the height of a person works out to be 18 m tall, alarm bells should start to ring. It is possible that the answer is out by a factor of ten and might really be 1.8 m. A close check should reveal the problem. Common sense plays a role in checking.

In your group, discuss each scenario and decide whether the following answers make sense. Report back to the class

A painter worked out that 4.376231 gallons of paint was required to paint a room. Explain what you think of that.

When calculating the number of buses to organize for a school trip, the teacher calculated the school would require 50 buses, because 150 ÷ 30 is 50. Have you ever seen 50 buses parked end-to-end?

A boy who lives 9 miles from school claims he can ride to school in 6 minutes. Would you believe him?

Mom's car was running on empty so she stopped to fill the tank up. It cost $4 to fill. Does that sound right?

The bill for a cart-load of groceries to feed a family of four amounted to $847. Does that sound right?

When calculating the mean weight for students in the class, your group works out the answer to be 8.4 lb. One member of the group thinks you have made a mistake. How would you respond?

Mom worked out that you can get 80, 250 mL glasses of cool drink from a 2 L bottle but only 10, 200 mL glasses. There seems to be a problem here. What do you think it might be?

Dad called to order 30 cubic meters of sand to top-dress the lawn. When he came home that night he was shocked. Why?

When working out the mean height of the class your group calculates the answer to be 5.7 inches. Does this seem right?

Dad worked out the dosage of medicine needed as 50 mL and decided to check his calculation. Why?

A girl at your school claims to run the 100 m in 1.83 seconds. Would you believe her? What mistake do you think she might have made?

When you went to the shop to buy a bag of chips and a drink and a chocolate bar the shopkeeper wanted to charge you $16.40. What might you say?

Estimation and rounding

Front-end estimation

Teacher information

A simple form of estimation for adding a series of numbers is called front-end rounding. There are two important steps when using this technique.

1. Look at the left-most digit in a number.
2. Consider the place-value of the digit.

For example, when adding you would simply ignore all the digits and then make use of place value to produce an estimate. Later on students may be encouraged to consider the effect the 900 has on the estimate.

3	215
6	910
4	342

It should be noted children find this technique fairly simple to use as it simply involves covering the digits to the right of the left-most digit. As the children become more proficient in the use of this method they should be encouraged to look at the digits in the next place along in order to produce an estimate closer to the actual result. This technique relies on children being able to add numbers beyond the basic facts. There are several ways the front-end strategy may be varied, including clustering numbers according to their magnitudes; i.e. numbers in the hundreds, thousands, etc.

3 + 6 + 4 = 13
So my estimate would be 13,000

(Note: The front-end rounding technique will always produce an underestimate.)

Extension

Students can try estimating using rounding and compare the results from both methods.

Answers

1. (a) 280 (estimate less than real answer) 298 (real answer)
(b) 130 (estimate less than real answer) 155 (real answer)
(c) 120 (estimate less than real answer) 139 (real answer)
(d) 1,300 (estimate less than real answer); 1,462 (real answer)
(e) 1,400 (estimate less than real answer); 1,552 (real answer)
(f) 2,600 (estimate less than real answer); 2,727 (real answer)
(g) 10,000 (estimate less than real answer); 12,048 (real answer)
(h) 11,000 (estimate less than real answer); 12,867 (real answer)
(i) 22,000 (estimate less than real answer); 23,088 (real answer)
The front-end method of estimation produces an underestimate.
(j) 6,000 (estimate less than real answer); 8,424 (real answer)
Note. Even though 980 is almost 1,000 this does not count when using a front-end estimate.
(k) 8,000 (estimate less than real answer); 11,501 (real answer)
The larger the last three digits the less accurate the front-end method becomes.
(l) 10,000 (estimate less than real answer); 10,500 (real answer)
The smaller the last three digits the closer the front-end method is to the exact answer.

2. Answers will vary. Front-end estimation will always produce an estimate below the actual result.

Front-end estimation

Front-end estimation is a simple way to estimate the answer to an addition calculation. All you do is look at the left-most digits of the calculation and use them to form your estimate. For example, to estimate the result of adding ...

7,745
2,347
4,678
+ 1,398

... you can cover all the digits to the right of the thousands digits and add the remaining digits. This would roughly tell you the size of your answers in thousands.

7,745
2,347
4,678
+ 1,398

7 + 2 + 4 + 1 = 14
so I would estimate the answer to be about 14,000.

Compare all your estimates to the real answers.
What do you notice?

1. Try using front-end estimation for each of the following calculations.

(a)	(d)	(g)	(j)
79	122	2,958	1,493
82	347	1,602	980
64	592	4,007	2,206
+ 73	+ 401	+ 3,481	+ 3,745

(b)	(e)	(h)	(k)
36	106	1,951	2,987
58	299	3,096	3,851
49	310	4,152	1,766
+ 12	+ 837	+ 3,668	+ 2,897

(c)	(f)	(i)	(l)
45	416	9,154	2,074
37	932	2,068	3,109
16	871	3,607	4,206
+ 41	+ 508	+ 8,259	+ 1,111

2. How accurate do you think front-end estimation is? Write about your findings.

Estimation and rounding

Rounding to estimate (subtraction)

Teacher information

In this activity the students are introduced to the idea of rounding numbers so an estimate may be formed. Initially they are encouraged to round to the nearest 100 or 1,000 so they are only required to calculate with a single digit.

It is important that the estimation process be quick and simple otherwise it will not be used. Later in the activity the students should be encouraged to discuss the results of their estimates and whether rounding to the nearest 10 may have produced an estimate closer to the answer.

There will always be a trade-off between the speed or ease of the estimate and its accuracy. Remember, it is better students make an estimate, than not make one at all.

Answers

Note. Actual answer is in brackets.

1. (a) 600 – 200 = 400 (337)
   (b) 400 – 300 = 100 (152)
   (c) 600 – 300 = 300 (283)
   (d) 400 – 200 = 200 (246)
   (e) 1,700 – 400 = 1,300 (1,353)
   (f) 2,600 – 1,400 = 1,200 (1,287)
   (g) 3,000 – 2,000 = 1,000 (606)
   (h) 4,100 – 4,000 = 100 (153)
   (i) 6,700 – 2,200 = 4,500 (4,548)
   (j) 8,200 – 6,000 = 2,200 (2,282)
   (k) 7,500 – 5,500 = 2,000 (1,955)
   (l) 9,600 – 7,300 = 2,300 (2,288)

Note. 1. (d), (e) and (f) provide opportunity for discussion. In 1.(d) rounding to the nearest 100 produces an estimate of 200, whereas rounding to the nearest 10 produces an estimate of 250 which is closer to the exact answer. The purpose behind making the estimate will dictate which is the more useful estimate. In 1.(e) the decision to round to the nearest 100 was made because the smaller of the two numbers was only in the hundreds.

2. *Teacher check*

Rounding to estimate (subtraction)

Rounding may be used to help make estimates.

1. *Consider the following subtraction.*

$$\begin{array}{r} 7{,}745 \\ -\ 2{,}347 \end{array}$$

2. *Round the numbers to the nearest 1,000.*

$$\begin{array}{r} 8{,}000 \\ -\ 2{,}000 \\ \hline 6{,}000 \end{array}$$

3. *This produces an estimate of 6,000. The exact answer is 5,398. Notice what happens if the numbers are rounded to the nearest hundred.*

$$\begin{array}{r} 7{,}700 \\ -\ 2{,}300 \\ \hline 5{,}400 \end{array}$$

Rounding to the nearest hundred is much closer than the estimate produced by rounding to the nearest thousand.

1. Use rounding to work out an estimate for each of the following. You will need to decide whether you should round to the nearest 100 or 1,000. When you have completed your estimates, calculate the exact answer and write it in the box below.

(a) 574 − 237

(b) 439 − 287

(c) 551 − 268

(d) 396 − 150

(e) 1,724 − 371

(f) 2,649 − 1,362

(g) 2,954 − 2,348

(h) 4,106 − 3,953

(i) 6,706 − 2,158

(j) 8,244 − 5,962

(k) 7,493 − 5,538

(l) 9,564 − 7,276

2. How do you think rounding to estimate an answer could help you to check your work?

____

____

Estimation and rounding

Rounding to estimate (multiplication)

Teacher information

Both front-end and rounding techniques may be used to form an estimate when multiplying numbers; however, rounding tends to provide a closer estimate. Even rounding can be imprecise. For example, when multiplying two numbers, if both are rounded down, the estimate will be below the actual answer. Students should be encouraged to consider how the rounding process may have affected the estimate and then adjust the estimate accordingly.

Rounding to the nearest ten, rather than hundred, will also have a bearing on the estimate that is formed. Some students will find it difficult to estimate using more than single digits so do not pressure them to round to the nearest ten, even though it might give a more accurate estimate. The students should simply be encouraged to make an estimate.

Extension

Draw the student's attention to Questions 12, 13 and 14. The estimate is the same. Discuss.

The students should also compare the use of front-end and rounding techniques for estimating the answer to the first ten questions. They should find in most cases rounding gives a closer estimate.

Comparison of front-end and rounding techniques

	Front-end method	*Real answer*	*Rounding technique*	Comparison
1.	60 x 30 = 1,800	63 x 34 = 2,142	60 x 30 = 1,800	same
2.	50 x 40 = 2,000	2,419	2,400	closer
3.	70 x 30 = 2,100	2,812	3,200	closer
4.	70 x 30 = 2,100	2,686	2,400	closer
5.	80 x 40 = 3,200	3,690	4,000	closer
6.	60 x 70 = 4,200	4,686	4,900	closer
7.	90 x 40 = 3,600	3,906	3,600	same
8.	90 x 30 = 2,700	3,384	3,600	closer
9.	90 x 40 = 3,600	4,180	4,000	closer
10.	80 x 60 = 4,800	5,355	5,400	closer

Answers

1. *60 x 30 = 1,800*
2. *60 x 40 = 2,400*
3. *80 x 40 = 3,200*
4. *80 x 30 = 2,400*
5. *80 x 50 = 4,000*
6. *70 x 70 = 4,900*
7. *90 x 40 = 3,600*
8. *90 x 40 = 3,600*
9. *100 x 40 = 4,000*
10. *90 x 60 = 5,400*
11. *120 x 30 = 3,600*
12. *130 x 40 = 5,200*
13. *130 x 40 = 5,200*
14. *130 x 40 = 5,200*
15. *250 x 80 = 20,000*
16. *230 x 140 = 32,200*
17. *280 x 130 = 36,400*
18. *220 x 270 = 59,400*
19. *290 x 220 = 63,800*
20. *390 x 490 = 191,100*

While rounding to the nearest ten, before multiplying, produces an estimate closer to the exact answer it is much more cumbersome and time consuming, and therefore children are less likely to make an estimate.

Rounding to estimate (multiplication)

Rounding may be used to estimate the result of multiplying two numbers.

- *Consider the following multiplication.*

74
x 47

- *Round the numbers to the nearest 10.*

70
x 50
3,500

- *This produces an estimate of 3,500. The exact answer is 3,478. Notice what happens if a front-end strategy is used.*

70
x 40
2,800

The resulting answer is much further away from the exact answer!

I would do 7 x 5 is 35 and then adjust for the fact that I was really multiplying by tens.

You may like to compare rounding to the nearest ten with rounding to the nearest 100.

Try using rounding to work out an estimate for each of the following.

1. 63 x 34	6. 66 x 71	11. 117 x 32	16. 230 x 142
2. 59 x 41	7. 93 x 42	12. 132 x 37	17. 278 x 131
3. 76 x 37	8. 94 x 36	13. 129 x 35	18. 218 x 267
4. 79 x 34	9. 95 x 44	14. 134 x 39	19. 291 x 216
5. 82 x 45	10. 85 x 63	15. 254 x 82	20. 389 x 491

Estimation and rounding

Multiplying

The activity is designed to help students see the impact of the rounding process on the forming of an estimate. This in turn will assist them to make more accurate estimates. For example, when multiplying, if both numbers are rounded down the estimate will be well below the actual answer. The student may choose to adjust the estimate up to compensate or simply remember that the answer should be above the estimate

Answers

1.

Question	Round both numbers up	Round both numbers down	Round 1st number up and the 2nd down	Round 1st number down and the 2nd up	Exact answer
3.41 x 5.92	4 x 6 = 24	3 x 5 = 15	4 x 5 = 20	3 x 6 = 18	20.1872
8.27 x 3.16	9 x 4 = 36	8 x 3 = 24	9 x 3 = 27	8 x 4 = 32	26.1332
5.32 x 8.57	6 x 9 = 54	5 x 8 = 40	6 x 8 = 48	5 x 9 = 45	45.5924
4.13 x 5.93	5 x 6 = 30	4 x 5 = 20	5 x 5 = 25	4 x 6 = 24	24.4909
9.78 x 8.37	10 x 9 = 90	9 x 8 = 72	10 x 8 = 80	9 x 9 -= 81	81.8586
3.13 x 4.25	4 x 5 = 20	3 x 4 = 12	4 x 4 = 16	3 x 5 = 15	13.3025
4.86 x 6.93	5 x 7 = 35	4 x 6 = 24	5 x 6 = 30	4 x 7 = 28	33.6798
5.17 x 6.21	6 x 7 = 42	5 x 6 = 30	6 x 6 = 36	5 x 7 = 35	32.1057
4.51 x 8.54	5 x 9 = 45	4 x 8 = 32	5 x 8 = 40	4 x 9 = 36	38.5154
2.89 x 7.87	3 x 8 = 24	2 x 7 = 14	3 x 7 = 21	2 x 8 = 16	22.7443

2. *Teacher check*
3. *Some forms of rounding are closer to the exact answer.*
4. *The result is larger than the exact answer.*
5. *The result is smaller than the exact answer.*
6. *The result is a little closer to the exact answer and may be above or below the exact result.*
7. *Teacher check*
8. *Teacher check*

1. Complete the following table by rounding the numbers to the nearest whole numbers in different ways.

Question	*Round both numbers up*	*Round both numbers down*	*Round 1st number up and the 2nd down*	*Round 1st number down and the 2nd up*	*Exact answer*
3.41 x 5.92	4 x 6 = 24	3 x 5 = 15	4 x 5 = 20	3 x 6 = 18	20.1872
8.27 x 3.16					
5.32 x 8.57					
4.13 x 5.93					
9.78 x 8.37					
3.13 x 4.25					
4.86 x 6.93					
5.17 x 6.21					
4.51 x 8.54					
2.89 x 7.87					

2. Color the results that are closest to the exact number.

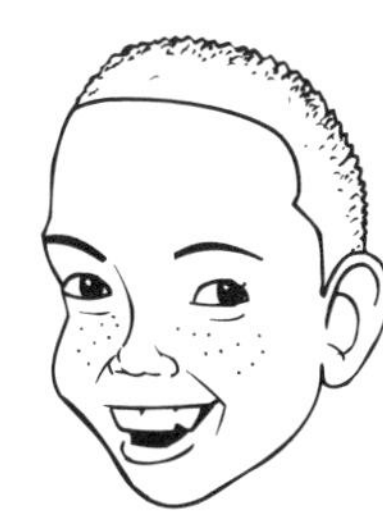

3. What do you notice about the results?

____

____

____

4. When you round both numbers up, is the result smaller or larger than the exact answer?

____

____

____

5. When you round both numbers down, is the result smaller or larger than the exact number?

____

____

6. What happens when you round one number up and the other down?

____

____

____

7. If you know your estimate will be below the real answer, what might you do?

____

____

____

____

8. If you know your estimate will be above the real answer, what might you do?

____

____

____

Estimation and rounding

Estimating with division

There are few different methods that may be used to estimate the result of division.

Method 1

Focus on the leading digits.

Example $4\overline{)13{,}244}$ (quotient begins 3 —)

A reasonable estimate would be that the answer should be in the 3000s.

Method 2

Adjust the divisor and the dividend so they are compatible; that is, the division is made simple.

Example $4\overline{)13{,}244}$ *becomes* $4\overline{)12{,}000}$

This technique relies on a knowledge of basic number facts.

Answers

1. (a) $4\overline{)827}$ *close to* $800 \div 4$
   estimate 200

   (b) $8\overline{)9{,}147}$ *close to* $8{,}000 \div 8$
   estimate 1,000

   (c) $6\overline{)723}$ *close to* $600 \div 6$
   estimate 100

   (d) $7\overline{)7{,}392}$ *close to* $7{,}000 \div 7$
   estimate 1,000

   In each case the estimate is an underestimate and may be adjusted up.

2. (a) $5\overline{)26{,}168}$
   would be easy if $25{,}000 \div 5$
   estimate 5,000

   (b) $7\overline{)22{,}644}$
   would be easy if $21{,}000 \div 7$
   estimate 3,000

   (c) $8\overline{)17{,}601}$
   would be easy if $16{,}000 \div 8$
   estimate 2,000

   (d) $4\overline{)25{,}542}$
   would be easy if $24{,}000 \div 4$
   estimate 6,000

   In each case the estimate is an underestimate and may be adjusted up.

Estimating with division

When estimating the result of a division you can just focus on the left-most digits.

For example, an estimate for 7)23,641 may be made by thinking ...

23 ÷ 7 is close to 21 ÷ 7, therefore, a reasonable estimate is 3,000.

1. Try estimating the answer for these problems using this method. As you try each, explain your thinking.

(a) 4)827 ____________________

(d) 7)7,392 ____________________

(b) 8)9,147 ____________________

(c) 6)723 ____________________

Another way of estimating the result of a division is outlined below.

4)33,179 —Think 4)32,000 would be easier. Estimate 8,000.

2. Try estimating the answer for these problems using this method. As you try each one, explain your thinking.

(a) 5)26,168 ____________________

(c) 8)17,601 ____________________

(b) 7)22,644 ____________________

(d) 4)25,542 ____________________

Estimation and rounding

Placing the point

Teacher information

When multiplying the numbers involving decimals students are often taught to count the number of decimal places in order to insert the decimal point in the correct spot. This does little to assist the development of number sense. Rather, it would be preferable for students to form an estimate prior to performing the calculation and then insert the decimal point based on the estimate.

For example, 27.6 x 31.24 should produce an answer in the hundreds, not the tens or thousands. An order of magnitude check would determine this and allow for the insertion of a decimal point. More sophisticated checking techniques would indicate that the result was greater than 600 (front-end) and probably closer to 900 (30 x 30 – using rounding).

Multiplying 276 by 3,124 produces 862,224.

Having made an estimate that the answer is in the hundreds, specifically between 600 and 900 allows for the decimal point to be placed,

862.224

Hit the target!

Rules

1. *Players choose a target number or the teacher writes some target numbers on the chalkboard from which the players choose one (e.g. 545).*
2. *One player chooses a number and enters it into the calculator and presses the multiplication key (e.g. 27 x).*
3. *The second player then has five seconds to enter a number into the calculator and press "=". If the result is within the target range (for example ±5) then the second player wins. If the result is outside the target range the first player then has a turn at trying to reach the target, using the number shown on the display.*

Player 1   27 x 23 = 621   (too high)
Player 2   621 x 0.9 = 558.69   (close)
Player 1   558.9 x 0.98 = 547.722   (close enough)

Variation

- *Instead of using the number shown on the display, the opposing player chooses the number that is entered.*

  Player 1   27 x 23 = 621   (too high)
  Player 1 enters 16 into the calculator
  Player 2   16 x 34 = 544   (close enough)

- *Players could be awarded one point for being the closest. Play could continue until one player reaches 10 points.*

Answers

1. (a) 272.34
(b) 193.12
(c) 121.52
(d) 60.03
(e) 17.48
(f) 15.54
(g) 5.52
(h) 5.76
(i) 11
(j) 0.72
(k) 0.48
(l) 1.28
(m) 2.42
(n) 6.37
(o) 7.77
(p) 155.12
(q) 333.27
(r) 89.91

2. Teacher check

Placing the point

1. The following calculations have been completed but the decimal points have not been inserted. Use estimation techniques to help you place the decimal points.

(a)	15.3 x 17.8 27234	(g)	0.6 x 9.2 552	(m)	1.1 x 2.2 242
(b)	14.2 x 13.6 19312	(h)	0.9 x 6.4 576	(n)	1.3 x 4.9 637
(c)	12.4 x 9.8 12152	(i)	0.2 x 55 11	(o)	2.1 x 3.7 777
(d)	8.7 x 6.9 6003	(j)	0.9 x 0.8 72	(p)	221.6 x 0.7 15512
(e)	2.3 x 7.6 1748	(k)	0.6 x 0.8 48	(q)	144.9 x 2.3 33327
(f)	4.2 x 3.7 1554	(l)	1.6 x 0.8 128	(r)	99.9 x 0.9 8991

2. Describe how you decide where the decimal point should go in answers to multiplication sums like these.

____

____

____

____

____

____

____

____

Hit the target!

Ask the teacher or friend to write target numbers on the board (any numbers). You and your partner choose a target number.

Player 1 enters a number into the calculator (different from the target number) and presses the multiplication key **X**. *Player 2 is given the calculator. He/She must enter a number into the calculator and press equals* **=**. *If the result is close to the target number, player 2 gets the point. If it is not, player 1 gets the point.*

Estimation and rounding

The rounding game

Teacher information

Rounding is a process which is used to make the numbers being worked with simpler to use.

For example, multiplying 40 by 30 requires less "cognitive load" or processing than 43 x 34 because there is less to hold in your head or to write down.

The rounding game is designed to give children practice in rounding a number to a desired level of accuracy. Rounding is a prerequisite skill for many forms of estimation; therefore, children need to be secure in their ability to round.

Answers

Answers will vary.

The rounding game

A game for two to four players.

Materials

- A blank die or wooden cube marked 10, 100, or 1,000 on two sides. Or a spinner marked similarly.
- Number cards (below).

Rules

1. Players take turns to roll the die. The number shown on the die indicates the place to which the number is rounded.
2. After the die is rolled, the player picks up a card and rounds the number shown to the nearest 10, 100, or 1,000, depending on what was rolled on the die.
3. If the rounding is correct, the player scores a point. Play continues for several rounds, with the player who scores the most named as the winner.
4. If the other player is unsure, a teacher or third party can check if the rounding is correct.

For example, the die is rolled and lands with 1,000 showing on the uppermost face. A card is picked up showing the number 7,268 and the player rounds the number to the nearest 1,000—in this case, 7,000.

6,283	2,345	3,698	4,545	9,679	8,236	5,555	2,199	2,754	2,683
3,496	4,189	4,332	4,547	1,001	1,010	9,999	9,090	3,035	8,998
3,211	1,711	984	951	6,243	6,499	7,253	7,888	8,010	8,412
9,671	17,601	14,523	11,173	12,902	12,064	12,500	11,999	11,765	11,050
11,668	13,456	13,211	13,511	14,606	14,482	15,743	15,099	15,199	15,499
16,276	17,841	18,973	18,379	18,475	19,999	19,444	19,500	20,100	1,500

CALCULATORS

All the checking methods described in this book may be used to check the results of a calculation completed with the aid of a calculator. There are, however, some typical calculator errors that may be detected using some simple checking techniques. These techniques are described in this section.

Calculators

Pushing the wrong buttons

Teacher information

When using a calculator, mistakes may occur when pressing the keys.

Typically these include:

- *entering a digit twice (double pressing)*
- *omitting a digit*
- *changing the order of the digits*
- *keying in the wrong numbers/operation*

The first two mistakes may be picked up by an *order of magnitude* check. Prior to starting the calculation the student should be encouraged to consider whether the answer is likely to be in the hundreds, thousands, millions, etc. Entering a digit twice will increase the answer by a power of ten. Omitting a digit will reduce the answer by a power of ten. The main focus of *Pushing the wrong buttons* is on entering too many digits or not enough. For the most part *order of magnitude* checks will help to identify these errors.

Answers

1. (a) *Answer way too small.*
   (b) *Pressed "+" instead of "–"*
2. (a) *Should expect a larger answer.*
   (b) *Pressed "–" instead of "+"*
3. (a) *Answer way too small.*
   (b) *Pressed "÷" instead of "x"*
4. (a) *Too small. Should end in 8 not 5.*
   (b) *678 + 37—forgot to press 0*
5. (a) *Two-digit multiplication likely to give answer in hundreds or thousands—larger number most likely in thousands.*
   (b) *Possibly pressed 79 x 4*
6. (a) *Yes*
   (b) *1000 ÷ 20 is 50*
   (c) *990 ÷ 8 instead of 18*
7. *Answers will vary.*

Pushing the wrong buttons

Sometimes when using a calculator you might push the wrong button. In the following examples, some students have accidentally pressed the wrong buttons on their calculators. What do you think they did wrong?

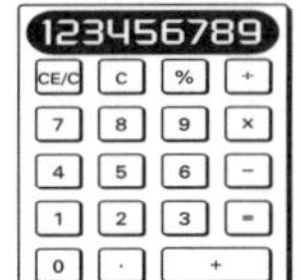

1. (a) Daniel meant to press 16 x 24 but the display showed 40, which he thought possibly couldn't be right. How did he know he wasn't right?

   ____

   (b) What buttons do you think he pressed?

   ____

2. (a) Susie thought she had entered 6,345 + 776 into her calculator but when the display showed 5,569 she looked puzzled. Why did she look puzzled?

   ____

   (b) What buttons might she have pressed?

   ____

3. (a) Soula was confused when 39 was shown on the display of her calculator after she had entered 975 x 25. Why was she confused?

   ____

   (b) What buttons might she have pressed?

   ____

4. (a) When Adrian entered 678 + 370 into his calculator the display showed 715. He thought that couldn't possibly be right. Why did he think the answer was wrong?

   ____

   ____

   (b) What buttons might he have pressed?

   ____

5. (a) Greg was amazed when his calculator showed an answer in the hundreds instead of the thousands when he entered 79 x 40 into the calculator. Why do you think he was expecting the answer to be in the thousands? Was he thinking correct?

   ____

   ____

   (b) What do you think happened to get an answer in the hundreds?

   ____

6. (a) When dividing 990 by 18 Louisa was expecting an answer of roughly 50, so when she got an answer of 123.75 she became concerned. Was she right to expect an answer near 50?

   ____

   ____

   (b) Explain why she may have reasoned this way.

   ____

   ____

   (c) What do you think happened when she pressed the buttons?

   ____

7. Make up a wrong button question for your partner to solve.

Calculators

Calculator errors

Teacher information

Calculator errors focuses on a range of typical keying-in errors including:

- *changing the order of the digits*
- *entering the wrong operation or numbers*
- *entering an extra digit*
- *omitting a digit*

For the most part *order of magnitude* checks will help identify if a mistake has been made.

When dealing with keying-in errors, students should be encouraged to examine the way their calculator works. For example, if a student realizes a keying-in-error has been made during a calculation, there may be no need to re-enter the calculation.

This demonstrates that even if all the correct keys are pressed some calculators will produce an incorrect result. You should know whether the calculator you commonly use gives the mathematically correct result or whether it gives the wrong answer. Whenever you pick up a calculator for the first time, try 7 + 3 x 5 to see whether it gives the correct result. Don't worry if your calculator gives an incorrect result as the only time this will be a problem is when more than one operation is used in a number sentence. If your calculator does not handle this properly you will need to change the order in which the calculation is done before entering it into the calculator.

Answers

1.

Calculation	Answer	Mistake	Correct answer	Check
23 x 27	*50*	*Pressed "+" instead of "x"*	*621*	*621 ÷ 23 should give 27. 50 ÷ 23 does not give 27*
45 x 67	*270*	*Pressed 46 x 6*	*3,015*	*50 x 70 = 3,500*
279 + 468	*279,468*	*Forgot to press "+"*	*1,076*	*Should give answer in hundreds*
931 + 797	*742,007*	*Pressed "x" instead of "+"*	*1,728*	*Expect answer in low thousands*
171 x 224	*159,264*	*Pressed 711 instead of 171*	*38,304*	*Expect answer in tens, not hundreds of thousands*
2,929 ÷ 20.2	*14.5*	*Pressed 202 instead of 20.2*	*145*	*Expect answer around 150 3,000 ÷ 20*
945 ÷ 27	*918*	*Pressed "–" instead of "÷"*	*35*	*Expect much smaller answer*
$^1/_2$ of 968	*1,936*	*Doubled instead of halved*	*484*	*Should be smaller rather than larger*
77 x 99	*76,923*	*Pressed 999 instead of 99*	*7,623*	*Answer too large. Should expect answer in thousands*
7 + 3 x 5	*50*	*Rule of order not applied. Did 7 + 3 = 10→ 10 x 5 = 50*	*22*	*Mental check*

Calculator errors

The following calculations were performed on a calculator. Unfortunately, the person pushing the buttons was tired and made a lot of errors. Work through each of the calculations and describe what went wrong. Suggest a checking method that might be used to tell whether the answer was correct. The first one has been done for you.

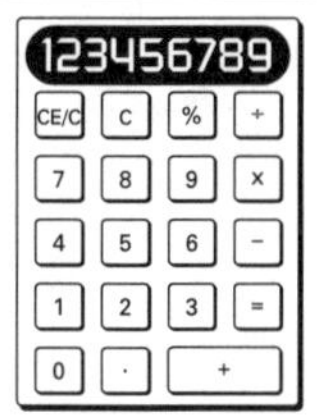

Calculation	Answer	Mistake	Correct answer	Check
23 x 27	50	Pressed "+" instead of "x"	621	621 ÷ 23 should give 27. 50 ÷ 23 does not give 27
45 x 67	270			
279 + 468	279,468			
931 + 797	742,007			
171 x 224	159,264			
2,929 ÷ 20.2	14.5			
945 ÷ 27	918			
$^{1}/_{2}$ of 968	1,936			
77 x 99	76,923			

This last one is a little tricky. Some calculators are programmed to give the correct answer of 22 while others will give the wrong answer of 50.

Find some different calculators in your classroom and try them to see which give the correct result and which give an incorrect result.

CASTING OUT NINES

Casting out nines is an old method of checking a calculation. While it is not used much in the calculator age, it will be of interest to more able students. It is based on observing patterns related to nine—hence the name of the technique, casting out nines.

Students will learn to find the digit sum of a number and use it to check the results of a calculation. The digit sum is simple to calculate – all you need to do is add the digits of a number until a single digit is formed; e.g. 629 → 6 + 2 + 9 → 17 → 1 + 7 → 8, so the digit sum is 8.

Casting out nines

Casting out nines

Teacher information

Casting out nines was known to the Romans in the third century AD, was used as an error checking method by the Hindus and Arabs and came in to general use in the eleventh century. It was used to check both abacus-based calculation and written calculation.

It appears that the name *casting out nines* developed when people used early calculating devices such as the abacus. When working in base 10, beads were placed on to prongs of the abacus until there were ten on the prong. The ten beads would then be traded for one ten. Physically, this meant that nine beads were discarded (cast out) and one bead was placed onto the prong to the left of the previous prong on the abacus.

This principle is behind all trading games. This method of checking a calculation is rarely used today. Often it was taught using rote methods and children failed to understand the mathematical principles behind it. Even if children choose not to adopt it as a checking procedure, *casting out nines* may be considered as an example of one of the many fascinating patterns to be found in mathematics.

Once the process of *casting out nines* is mastered, students can become quite adept at using it to check calculations. The *casting out nines* method is closely related to the test for divisibility by nine. It is worthwhile reviewing this rule and the patterns in the "nine times table" before teaching students the *casting out nines* method.

Casting out nines involves combining the digits in a number until a single digit is produced. Comparing the single digit sum of the answer to the digit sums of the question will indicate whether the answer is incorrect.

Casting out nines will not necessarily show whether an answer is incorrect. Consider the following example.

When a number is divided by nine the remainder is closely related to the sum of the digits of the number. For example, when 72 is divided by 9 the remainder is 0. When the digit sum is found (7 + 2 = 9) and you "cast out" the nine (9 – 9 = 0) the answer matches the remainder.

Consider another example.

47 ÷ 9 = 5r2. In this case, the remainder is 2 recurring. The digit sum (4 + 7 = 11) is 11, and when the nine is cast out (11 – 9 = 2) the remainders match. This principle may be applied to any number and is the basis behind the casting out nines *checking method. This method is particularly useful for checking the results of addition, subtraction and multiplication, but becomes a little cumbersome when checking division.*

Specific teacher information

This activity reviews previous work on digit sums and the rule for divisibility by nine. It also extends divisibility by nine to include the notion of remainders.

Answers

1. (a) 9
   (b) 18
   (c) 27
   (d) 36
   (e) 45
   (f) 54
   (g) 63
   (h) 72
   (i) 81
   (j) 90
   (k) All equal nine.
2. 76r3
3. (a) 657 → 6 + 5 + 7 → 18 → 1 + 8 → 9 (no remainder)
   (b) 848 → 8 + 4 + 8 → 20 → 2 + 0 → 2 (remainder = 2)
   (c) 514 → 5 + 1 + 4 → 10 → 1 + 0 → 1 (remainder = 1)
   (d) 748 → 7 + 4 + 8 → 19 → 1 + 9 → 10 → 1 + 0 → 1 (remainder = 1)
   (e) 432 → 4 + 3 + 2 → 9 (no remainder)
   (f) 329 → 3 + 2 + 9 → 14 → 1 + 4 → 5 (remainder = 5)
   (g) 666 → 6 + 6 + 6 → 18 → 1 + 8 → 9 (no remainder)
   (h) 1,018 → 1 + 0 + 1 + 8 → 10 → 1 + 0 → 1 (remainder = 1)
4. When the digit sum is nine, there will not be a remainder. A digit sum other than nine means there will be a remainder.

Casting out nines

1. Complete the nine times table.

   (a) $1 \times 9 =$ ______

   (b) $2 \times 9 =$ ______

   (c) $3 \times 9 =$ ______

   (d) $4 \times 9 =$ ______

   (e) $5 \times 9 =$ ______

   (f) $6 \times 9 =$ ______

   (g) $7 \times 9 =$ ______

   (h) $8 \times 9 =$ ______

   (i) $9 \times 9 =$ ______

   (j) $10 \times 9 =$ ______

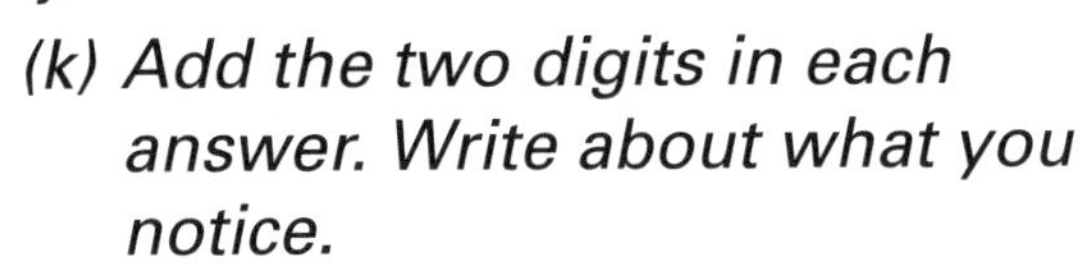

   (k) Add the two digits in each answer. Write about what you notice.

   ____

   ____

   ____

This is called the "casting out nines" method of checking calculations.

If a number is exactly divisible by nine (i.e. there is no remainder), then its digits should add up to nine.

For example:

891 is exactly divisible by 9 because

$$8+9+1 \rightarrow 18 \rightarrow 1+8 \rightarrow 9$$

687 is not divisible by 9 because

$$6+8+7 \rightarrow 21 \rightarrow 2+1 \rightarrow 3$$

2. Try dividing 9 into 687. Write down the remainder.

3. Use the "cast out" method to determine whether the following numbers are divisible by 9 without leaving a remainder, then write down the value when all the digits are added.

   (a) 657 ______________________________

   (b) 848 ______________________________

   (c) 514 ______________________________

   (d) 748 ______________________________

   (e) 432 ______________________________

   (f) 329 ______________________________

   (g) 666 ______________________________

   (h) 1,018 ______________________________

4. Write about what you notice.

   ____

   ____

   ____

   ____

   ____

   ____

Casting out nines

Addition (+)

Teacher information

This activity builds on previous work with digit sums. It should be noted that this method is not foolproof. If the digit sums do not match, the answer is wrong. If the digit sums match, it is likely that the answer is correct, although it is possible it may be wrong (see examples below).

Answers

(a)
527 → 5 + 2 + 7 → 14 → 5
+ 258 → 2 + 5 + 8 → 15 → 6 } 5 + 6 → 11 → **2**
775 → 7 + 7 + 5 → 19 → 10 → **1**

The two do not match so the answer is wrong—should be 785.

(b)
144 → 1 + 4 + 4 → 9
+ 717 → 7 + 1 + 7 → 15 → 6 } 9 + 6 → 15 → **6**
861 → 8 + 6 + 1 → 15 → **6**

The two match so the answer is probably correct.

(c)
462 → 4 + 6 + 2 → 12 → 3
+ 349 → 3 + 4 + 9 → 16 → 7 } 3 + 7 → 10 → **1**
801 → 8 + 0 + 1 → **9**

The two do not match so the answer is wrong—should be 811.

(d)
371 → 3 + 7 + 1 → 11 → 2
+ 243 → 2 + 4 + 3 → 9 } 2 + 9 → 11 → **2**
614 → 6 + 1 + 4 → 11 → **2**

The two match so the answer is probably correct.

(e)
576 → 5 + 7 + 6 → 18 → 9
+ 248 → 2 + 4 + 8 → 14 → 5 } 9 + 5 → 14 → **5**
914 → 9 + 1 + 4 → 14 → **5**

Here is an example where the "casting out nines" method fails. Even though the values match, the answer is wrong—should be 824. You will note that the digit sums for 914 and 824 are exactly the same.

Teaching point
Students should be made aware that while the "casting out" method works well most of the time, it is not foolproof.

(f)
256 → 2 + 5 + 6 → 13 → 4
+ 732 → 7 + 3 + 2 → 12 → 3 } 4 + 3 → **7**
888 → 8 + 8 + 8 → 24 → **6**

The two do not match; therefore, the answer is wrong—should be 988.

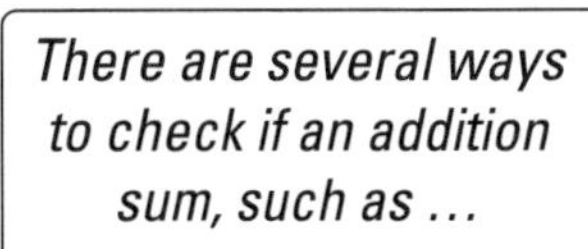

There are several ways to check if an addition sum, such as ...

357
+ 174
531

... is correct.

One way would be to reverse the calculation and add it again.

174
+ 357
531

You could subtract one number from the total to check.

531	531
– 357	– 174
174	357

Or you could try casting out nines.

To check a calculation such as ...

531
– 357
174

Simply add the digits in each of the numbers until a single digit remains.

357 → 3 + 5 + 7 → 15 → 1 + 5 → ❻
+ 174 → 1 + 7 + 4 → 12 → 1 + 2 → ❸
531 → 5 + 3 + 1 → 9

6 + 3 = 9 ✔

If the answer is correct the remainder of the answer will equal the sum of the remainders from the question.

415 → 4 + 1 + 5 → 10 → 1 + 0 → ❶
+ 282 → 2 + 8 + 2 → 12 → 1 + 2 → ❸
697 → 6 + 9 + 7 → 22 → 2 + 2 → 4

1 + 3 = 4 ✔

1. Use the "casting out nines" method to check whether the following problems are right or wrong.

(a) 527
+ 258
775

______ ◯ right ◯ wrong

(b) 144
+ 717
861

______ ◯ right ◯ wrong

(c) 462
+ 349
801

______ ◯ right ◯ wrong

(d) 371
+ 243
614

______ ◯ right ◯ wrong

(e) 576
+ 248
914

______ ◯ right ◯ wrong

2. Try making up some of your own.

Will casting out nines work with subtraction?

Try it and see!

Calculation checks

Subtraction (–)

Teacher information

The casting out nines method of subtraction is slightly more complicated than the method for addition but it is still worth considering.

Emphasize that when the digit sums match it is highly likely the answer is correct, but on the rare occasion the digits may be such that the answer is incorrect.

If, however, the digit sums do not match, the answer is definitely wrong.

Answers

1\. (a)

517 → ❹

– 76 → 4

441 → 9   4 + 9 → 13 → ❹

The two values match so the answer is probably correct.

(b)

676 → ❶

– 368 → 8

318 → 3   8 + 3 → 11 → ❷

The two values do not match so the answer is wrong—should be 308.

(c)

873 → ❾

– 491 → 5

372 → 3   5 + 3 → ❽

The two values do not match so the answer is wrong—should be 382.

(d)

2,417 → ❾

– 936 → 5

1,481 → 9   9 + 5 → 14 → ❺

The two values match so the answer is probably correct.

(e)

4,635 → ❾

– 2,998 → 1

1,647 → 9   1 + 9 → 10 → ❶

The two values do not match so the answer is wrong—should be 1,637.

(f)

1,078 → ❼

– 959 → 5

109 → 1   5 + 1 → ❻

The two values do not match so the answer is wrong—should be 119.

2\. (a)

26.2 → 2 + 6 + 2 → 10 → 1 + 0 → 1

– 19.2 → 1 + 9 + 2 → 12 → 1 + 2 → ❸

6.2 → 6 + 2 → ❽

8 + 3 → 11 → 1 + 1 → 2

The two values do not match so the answer is wrong—should be 7.

(b)

37.1 → 3 + 7 + 1 → 11 → 1 + 1 → 2

– 21.2 → 2 + 1 + 2 → ❺

15.9 → 1 + 5 + 9 → 15 → 1 + 5 → ❻

5 + 6 → 11 → 1 + 1 → 2

The two values match so the answer is probably correct.

3\.

666 → 6 + 6 + 6 → 18 → 1 + 8 ❾

– 555 → 5 + 5 + 5 → 15 → 1 + 5 ❻

444 → 4 + 4 + 4 → 12 → 1 + 2 3

The values match, which would seem to indicate the answer is correct. Clearly it is not. The casting out nines method will indicate when an answer is wrong but is not infallible when it comes to determining whether it is correct.

Subtraction (–)

The "casting out nines" method of checking a calculation works with subtraction as well as addition, but this time you compare the digit sums of the larger number with the digit sums of the smaller numbers.

For example:

4,361 → 4 + 3 + 6 + 1 → 14 → 1 + 4 → ❺
– 2,744 → 1 + 7 + 4 + 4 → 17 → 1 + 7 → 8
1,617 → 1 + 6 + 1 + 7 → 15 → 1 + 5 → 6
8 + 6 → 14 → 1 + 4 → ❺

The digit sums are the same, so it is likely the answer is correct.

If the two circled values are the same, the answer is probably correct.

1. Try using this technique to check these calculations.

(a) 517
– 76
441

(b) 676
– 368
318

(c) 873
– 491
372

(d) 2,417
– 936
1,481

(e) 4,635
– 2,998
1,647

(f) 1,078
– 959
109

I wonder if "casting out nines" works for decimal numbers?

26.2
– 19.2
6.2

37.1
– 21.2
15.9

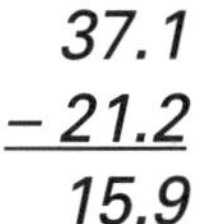

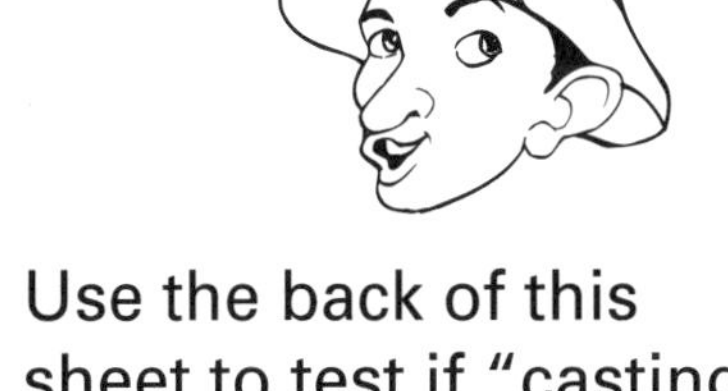

2. Use the back of this sheet to test if "casting out nines" works for decimal numbers.

3. The following calculation is obviously wrong, but what happens when you try "casting out nines"?

666
– 555
444

Use the back of this sheet to answer this question.

Casting out nines

Multiplication (x)

Teacher information

While the "casting out nines" method for multiplication is slightly more complicated than the addition and subtraction checks it is still worth considering. Once again it is important to emphasize that if the two values match it is highly likely that the answer is correct, but there may be rare cases when it isn't. If the two values are different, you can definitely say the answer is incorrect.

Answers

1\. (a)
37 → 3 + 7 → 10 → 1 + 0 → 1 } *1 x 7 →* **7**
x 16 → 1 + 6 → 7
592 → 5 + 9 + 2 → 16 → 1 + 6 → **7**
The two values match so the answer is probably correct.

(b)
54 → 5 + 4 → 9 } *9 x 9 → 81 → 8 + 1 →* **9**
x 36 → 3 + 6 → 9
1,934 → 1 + 9 + 3 + 4 → 17 → 1 + 7 → **8**
The two values do not match so the answer is wrong—should be 1,944.

(c)
91 → 9 + 1 → 10 → 1 + 0 → 1 } *1 x 8 →* **8**
x 62 → 6 + 2 → 8
5,662 → 5 + 6 + 6 + 2 → 19 → 1 + 9 → 10 → 1 + 0 → **1**
The two values do not match so the answer is wrong—should be 5,642.

(d)
67 → 6 + 7 → 13 → 1 + 3 → 4 } *4 x 1 →* **4**
x 73 → 7 + 3 → 10 → 1 + 0 → 1
4,891 → 4 + 8 + 9 + 1 → 22 → 2 + 2 → **4**
The two values match so the answer is probably correct.

(e)
83 → 8 + 3 → 11 → 1 + 1 → 2 } *2 x 4 →* **8**
x 67 → 6 + 7 → 13 → 1 + 3 → 4
5,561 → 5 + 5 + 6 + 1 → 17 → 1 + 7 → **8**
The two values match so the answer is probably correct.

2\. (a)
6.4 → 6 + 4 → 10 → 1 + 0 → 1 } *1 x 6 →* **6**
x 3.3 → 3 + 3 → 6
22.12 → 2 + 2 + 1 + 2 → **7**
The two values do not match so the answer is wrong—should be 21.12.

(b)
4.6 → 4 + 6 → 10 → 1 + 0 → 1 } *1 x 2 →* **2**
x 2.9 → 2 + 9 → 11 → 1 + 1 → 2
13.34 → 1 + 3 + 3 + 4 → 11 → 1 + 1 → **2**
The two values match so the answer is probably correct.

3\. (a)
32 → 3 + 2 → 5 } *5 x 1 →* **5**
x 100 → 1 + 0 + 0 → 1
500 → 5 + 0 + 0 → **5**
Even though the two values match, the answer quite clearly is wrong. The ***casting out nines*** *checking method will tell you when an answer is definitely wrong, but it is not infallible when it comes to the right answer—although in most cases it will work.*

The "casting out nines" method of checking a calculation may be used on a multiplication problem.

For example:

$$\begin{array}{r} 32 \rightarrow 3+2 \rightarrow 5 \\ \underline{\times\ 14} \rightarrow 1+4 \rightarrow 5 \end{array} \Big\} \ 5 \times 5 = 25 \rightarrow 2+5 \rightarrow \textcircled{7}$$

$$448 \rightarrow 4+4+8 \rightarrow 16 \rightarrow 1+6 \rightarrow \textcircled{7}$$

The remainders are multiplied, and a value is found by casting out nines. If this value and the remainder of the answer are the same, the calculation is probably correct.

For example:

$$\begin{array}{r} 71 \rightarrow 7+1 \rightarrow 8 \\ \underline{\times\ 89} \rightarrow 8+9 \rightarrow 17 \rightarrow 1+7 \rightarrow 8 \end{array} \Big\} \ 8 \times 8 = 64 \rightarrow 6+4 \rightarrow 1$$

$$6{,}319 \rightarrow 6+3+1+9 \rightarrow 19 \rightarrow 1+9 \rightarrow 10 \rightarrow 1+0 \rightarrow 1$$

Try checking these using the "casting out nines" method.

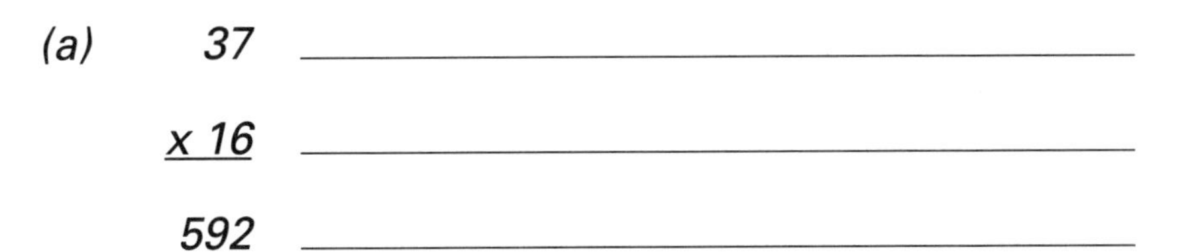

(a) 37 ______
x 16 ______
592 ______

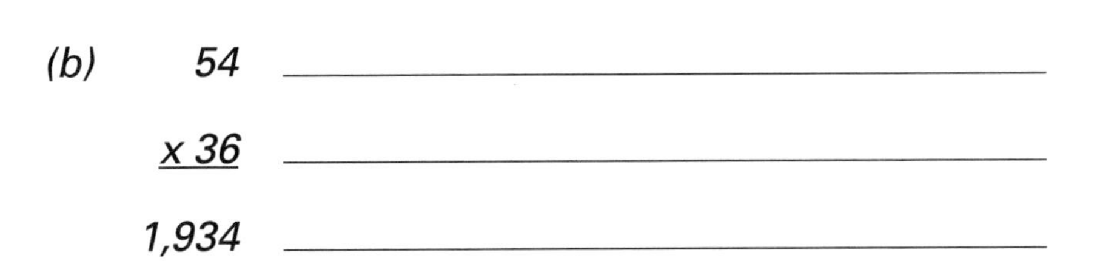

(b) 54 ______
x 36 ______
1,934 ______

(c) 91 ______
x 62 ______
5,662 ______

(d) 67 ______
x 73 ______
4,891 ______

(e) 83 ______
x 67 ______
5,561 ______

I wonder if "casting out nines" works when multiplying decimals?

(a) 6.4
x 3.3
22.12

(b) 4.6
x 2.9
13.34

2. Use the back of this sheet to test if "casting out nines" works when multiplying decimals.

The following calculation is obviously wrong, but what happens when you try "casting out nines"?

32
x 100
500

3. Use the back of this sheet to answer this question.

Casting out nines

Division (÷)

Teacher information

The "casting out nines" method for division is included for interest. It is quite involved and therefore other calculation checks would probably be more efficient.

Answers

(a) 38 → 3 + 8 → 11 → 1 + 1 → 2

7)266 *Divisor* → 7

Multiply the two values → 2 x 7 → 14

Cast out nines → 14 → 1 + 4 → ❺

Cast out nines for dividend → 266 → 2 + 6 + 6 → 14 → 1 + 4 → ❺

The two values match, so the answer is probably correct.

(b) 66 → 6 + 6 → 12 → 1 + 2 → 3

8)536 *Divisor* → 8

Multiply the two values → 3 x 8 → 24

Cast out nines → 24 → 2 + 4 → ❻

Cast out nines for dividend → 536 → 5 + 3 + 6 → 14 → 1 + 4 → ❺

The two values do not match, so the answer is wrong—should be 67.

(c) 84r3 → 8 + 4 → 12 → 1 + 2 → 3

6)506 *Divisor* → 6

Multiply the two values → 3 x 6 → 18

Add remainder → +3

Total → 21

Cast out nines → 21 → 2 + 1 → ❸

Cast out nines for dividend → 506 → 5 + 0 + 6 → 11 → 1 + 1 → ❷

The two values do not match, so the answer is wrong—should be 84r2.

(d) 93 → 9 + 3 → 12 → 1 + 2 → 3

12)1,116 *Divisor* → 12 → 1 + 2 → 3

Multiply the two values → 3 x 3 → ❾

Cast out nines → *No need to cast out nines in this case*

Cast out nines for dividend → 1116 → 1 + 1 + 1 + 6 → ❾

The two values match, so the answer is probably correct.

The "casting out nines" method may also be used to check a division sum. For example, consider a division without a reminder:

$13\overline{)494}$ = 38

(Cast out nines for answer)	3 + 8 = 11 → 1 + 1	→ 2
(Cast out nines for divisor)	13 → 1 + 3	→ 4
(Multiply)		→ 2 x 4 → ❽ ✔
(Cast out nines for the dividend)	494 → 4 + 9 + 4	→ 17 → 1 + 7 → ❽ ✔

If the two match, you are probably right.
If the division sum has a remainder, follow this procedure:

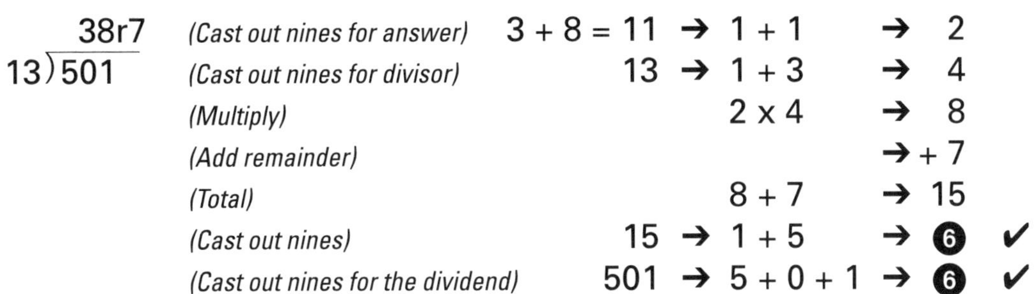

$13\overline{)501}$ = 38r7

(Cast out nines for answer)	3 + 8 = 11 → 1 + 1	→ 2
(Cast out nines for divisor)	13 → 1 + 3	→ 4
(Multiply)	2 x 4	→ 8
(Add remainder)		→ + 7
(Total)	8 + 7	→ 15
(Cast out nines)	15 → 1 + 5	→ ❻ ✔
(Cast out nines for the dividend)	501 → 5 + 0 + 1	→ ❻ ✔

If the two match, you are probably right. Try these:

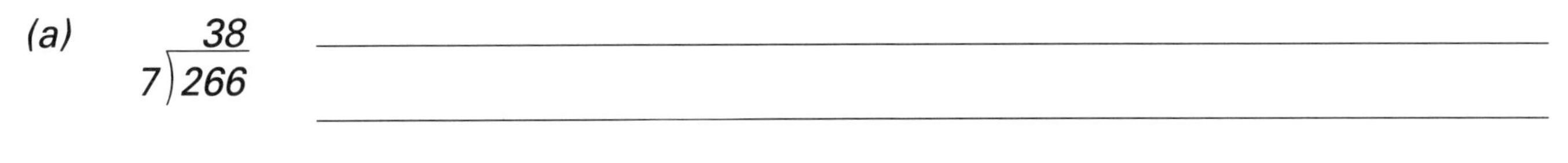

(a) $7\overline{)266}$ = 38

____

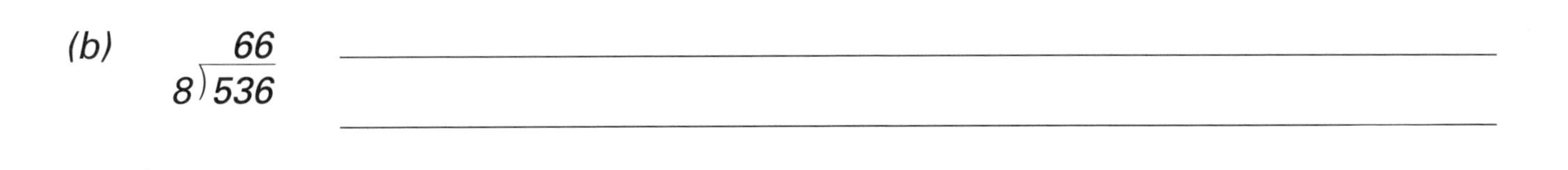

(b) $8\overline{)536}$ = 66

____

(c) $6\overline{)506}$ = 84r3

____

(d) $12\overline{)1,116}$ = 93

____

CHECKING CALCULATIONS A DIFFERENT WAY

Carrying out a calculation in a different way is an exact form of checking your work. There are several ways this can be done:

- Changing an addition into a subtraction (or vice versa; i.e. using an inverse operation).
- Changing a division into a multiplication (or vice versa).
- Altering the order of a calculation in the case of addition (but not subtraction). For example, you could check 7 + 2 + 3 by adding 7 and 3 and then 2. Likewise, the result of multiplying 3 and 8 may be checked by multiplying 8 by 3.
- Completing the same calculation using a different method. There are many alternative approaches that may be used.

Checking calculations a different way

Shopkeeper's method

The shopkeeper's method for counting back change was used prior to the introduction of modern cash registers. This was a real-life example of checking your calculation in another way. Essentially, the method involves counting back the change in increments. For example, if $50 was tendered to pay for goods to the value of $36.80, the change would be counted back in the following manner:

- *$36.80 and 20¢ makes $37;*
- *$37 and $3 makes $40;*
- *$40 and $10 makes $50.*

This may be illustrated on a number line.

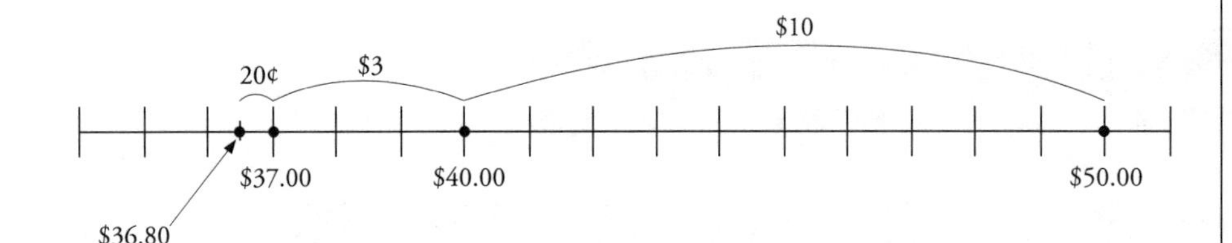

Answers

1. → *$12.10 and 90¢ makes $13*
   → *$13 and $7 makes $20*
   → *Change $7.90*
   → *Change given incorrect*

2. → *$22.45 and 5¢ makes $22.50*
   → *$22.50 and 50¢ makes $23*
   → *$23 and $7 makes $30*
   → *$30 and $10 makes $40*
   → *Correct change $17.50*
   → *Change given incorrect*

3. → *$1.65 and 5¢ makes $1.70*
   → *$1.70 and 30¢ makes $2*
   → *$2 and $8 makes $10*
   → *Correct change $8.35*
   → *Change given correct*

4. → *$21.70 and 30¢ makes $22*
   → *$22 and $8 makes $30*
   → *Correct change $8.30*
   → *Change given incorrect*

5. → *$35.35 and 5¢ makes $35.40*
   → *$35.40 and 60¢ makes $36*
   → *$36 and $4 makes $40*
   → *Correct change $4.65*
   → *Change given incorrect*

6. → *$39.10 and 90¢ makes $40*
   → *$40 and $10 makes $50*
   → *Correct change $11.90*
   → *Change given correct*

7. → *$7.65 and 5¢ makes $7.70*
   → *$7.70 and 30¢ makes $8*
   → *$8 and $2 makes $10*
   → *$10 and $40 makes $50*
   → *Correct change $42.35*
   → *Change given incorrect*

8. → *$15.40 and 60¢ makes $16*
   → *$16 and $4 makes $20*
   → *$20 and $30 makes $50*
   → *Correct change $34.60*
   → *Change given incorrect*

9. → *$61.80 and 60¢ makes $62*
   → *$62 and $8 makes $70*
   → *$70 and $30 makes $100*
   → *Correct change $38.20*
   → *Change given incorrect*

10. → *$3.45 and 5¢ makes $3.50*
    → *$3.50 and 50¢ makes $4*
    → *$4 and $6 makes $10*
    → *$10 and $90 makes $100*
    → *Correct change $96.55*
    → *Change given incorrect*

Shopkeeper's method

Long before electronic registers, shopkeepers would check the amount of change they were giving by counting back from the total to the amount tendered. This was really a checking approach based on repeating the calculation in a different way.

The shopkeeper's method involves "making up" the smaller number to the larger number. For example, if $50 was used to pay for $32.70 worth of goods, the shopkeeper would count the change back by stating:

$32.70 and 30¢ makes $33

$33 and $7 makes $40, and $10 makes $50

$17.30 in total is given in change. $32.70 + $17.30 = $50

Try using the shopkeeper's method to count back the change in each case. Check whether the amount given in change was correct.

1. Amount tendered $20
   Cost of goods $12.20
   Change given $8.90

   ____
   ____
   ____

   ◯ *correct* ◯ *incorrect*

2. Amount tendered $40
   Cost of goods $22.45
   Change given $17.65

   ____
   ____
   ____

   ◯ *correct* ◯ *incorrect*

3. Amount tendered $10
   Cost of goods $1.65
   Change given $8.35

   ____
   ____
   ____

   ◯ *correct* ◯ *incorrect*

4. Amount tendered $30
   Cost of goods $21.70
   Change given $18.30

   ____
   ____
   ____

   ◯ *correct* ◯ *incorrect*

5. Amount tendered $40
   Cost of goods $35.35
   Change given $4.55

   ____
   ____
   ____

   ◯ *correct* ◯ *incorrect*

6. Amount tendered $50
   Cost of goods $39.10
   Change given $10.90

   ____
   ____
   ____

   ◯ *correct* ◯ *incorrect*

7. Amount tendered $50
   Cost of goods $7.65
   Change given $33.35

   ____
   ____
   ____

   ◯ *correct* ◯ *incorrect*

8. Amount tendered $50
   Cost of goods $15.40
   Change given $35.60

   ____
   ____
   ____

   ◯ *correct* ◯ *incorrect*

9. Amount tendered .. $100
   Cost of goods $61.80
   Change given $39.20

   ____
   ____
   ____

   ◯ *correct* ◯ *incorrect*

10. Amount tendered .. $100
    Cost of goods $3.45
    Change given $98.55

    ____
    ____
    ____

    ◯ *correct* ◯ *incorrect*

Why are the benches at shops called counters?

Apparently, shopkeepers would place their counting boards (which were a little like the modern-day abacus) on the bench or tabletop and work out the cost of items and the change.

Checking calculations a different way

Addition

Teacher information

A calculation may be checked by using an inverse operation.

For example, the sum …

$$\begin{array}{r} 364 \\ +\,239 \\ \hline 603 \end{array}$$

may be checked using subtraction!

$$\begin{array}{r} 603 \\ -\,239 \\ \hline 364 \end{array} \qquad \begin{array}{r} 603 \\ -\,364 \\ \hline 239 \end{array}$$

A subtraction may be checked using addition.

$$\begin{array}{r} 364 \\ -\,127 \\ \hline 192 \end{array} \qquad \begin{array}{r} 192 \\ +\,127 \\ \hline 364 \end{array}$$

A multiplication may be checked using division, although in reality few people do so because they experience difficulty with division. It would probably be better to encourage them to use an alternative checking strategy.

A similar approach may be used for checking division but it is probably simpler to add the products of each multiplication, along with the remainder.

Answers

1. (a) $\begin{array}{r} 57 \\ +\,86 \\ \hline 143 \end{array} \rightarrow \begin{array}{r} 143 \\ -\,57 \\ \hline 86 \end{array}$ *or* $\begin{array}{r} 143 \\ -\,86 \\ \hline 57 \end{array}$

(b) $\begin{array}{r} 79 \\ +\,65 \\ \hline 144 \end{array} \rightarrow \begin{array}{r} 144 \\ -\,79 \\ \hline 65 \end{array}$ *or* $\begin{array}{r} 144 \\ -\,65 \\ \hline 79 \end{array}$

(c) $\begin{array}{r} 83 \\ +\,94 \\ \hline 177 \end{array} \rightarrow \begin{array}{r} 177 \\ -\,83 \\ \hline 94 \end{array}$ *or* $\begin{array}{r} 177 \\ -\,94 \\ \hline 83 \end{array}$

(d) $\begin{array}{r} 112 \\ +\,79 \\ \hline 191 \end{array} \rightarrow \begin{array}{r} 191 \\ -\,112 \\ \hline 79 \end{array}$ *or* $\begin{array}{r} 191 \\ -\,79 \\ \hline 112 \end{array}$

(e) $\begin{array}{r} 134 \\ +\,53 \\ \hline 187 \end{array} \rightarrow \begin{array}{r} 187 \\ -\,134 \\ \hline 53 \end{array}$ *or* $\begin{array}{r} 187 \\ -\,53 \\ \hline 134 \end{array}$

(f) $\begin{array}{r} 147 \\ +\,54 \\ \hline 201 \end{array} \rightarrow \begin{array}{r} 201 \\ -\,147 \\ \hline 54 \end{array}$ *or* $\begin{array}{r} 201 \\ -\,54 \\ \hline 147 \end{array}$

(g) $\begin{array}{r} 387 \\ +\,286 \\ \hline 575 \end{array} \rightarrow \begin{array}{r} 575 \\ -\,387 \\ \hline 286 \end{array}$ *or* $\begin{array}{r} 575 \\ -\,286 \\ \hline 387 \end{array}$

(h) $\begin{array}{r} 476 \\ +\,165 \\ \hline 641 \end{array} \rightarrow \begin{array}{r} 641 \\ -\,476 \\ \hline 165 \end{array}$ *or* $\begin{array}{r} 641 \\ -\,165 \\ \hline 476 \end{array}$

(i) $\begin{array}{r} 583 \\ +\,294 \\ \hline 877 \end{array} \rightarrow \begin{array}{r} 877 \\ -\,583 \\ \hline 294 \end{array}$ *or* $\begin{array}{r} 877 \\ -\,294 \\ \hline 583 \end{array}$

2. *Using the inverse operation—addition.*

(a) $\begin{array}{r} 164 \\ -\,43 \\ \hline 121 \end{array} \rightarrow \begin{array}{r} 121 \\ +\,43 \\ \hline 164 \end{array}$

(b) $\begin{array}{r} 379 \\ -\,265 \\ \hline 114 \end{array} \rightarrow \begin{array}{r} 114 \\ +\,265 \\ \hline 379 \end{array}$

(c) $\begin{array}{r} 483 \\ -\,294 \\ \hline 189 \end{array} \rightarrow \begin{array}{r} 189 \\ +\,294 \\ \hline 483 \end{array}$

If you are not sure whether the answer to a calculation is correct, you can repeat it. Instead of just doing the calculation in the same way, you can check an addition sum by using subtraction and multiplication by division.

To check	I did	I could also have done
45 + 67 = 112	112 − 45 = 67	112 − 67 = 45

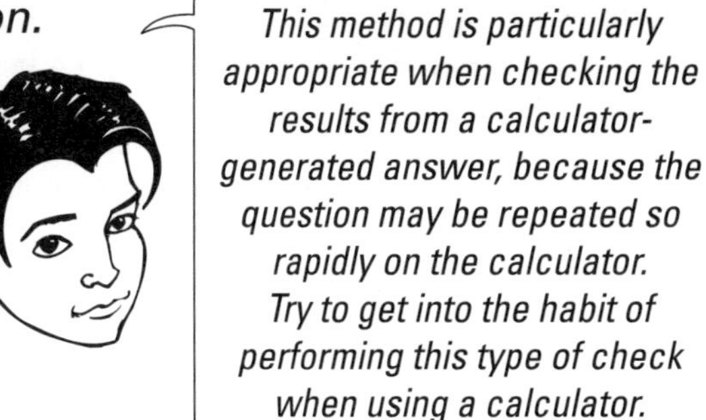

1. Complete the following calculations and then check using an inverse operation.

(a) 57 + 86 = 143 — check: 143 − 57 = 86; 143 − 86 = 57

(d) 112 + 79 — check

(g) 389 + 286 — check

(b) 79 + 65 — check

(e) 134 + 53 — check

(h) 476 + 165 — check

(c) 83 + 94 — check

(f) 147 + 54 — check

(i) 583 + 294 — check

2. Explain how you would check a subtraction.

____

____

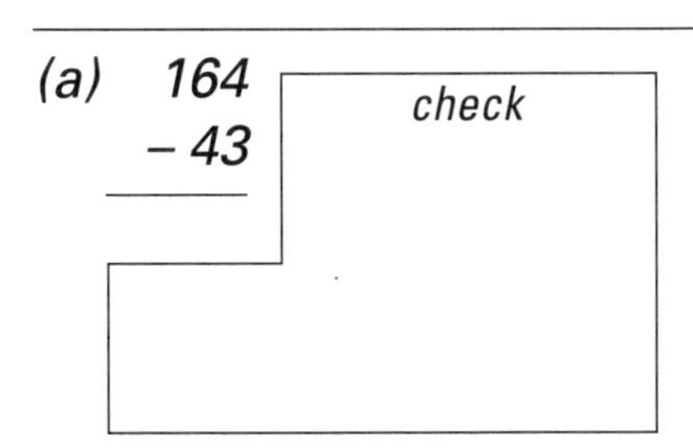

(a) 164 − 43 — check

(b) 379 − 265 — check

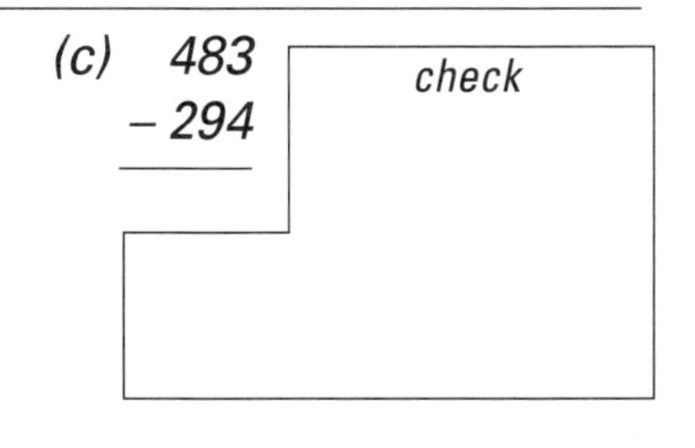

(c) 483 − 294 — check

Checking calculations a different way

Multiplication

Teacher information

A multiplication may be checked using division, although in reality few people do so because they experience difficulty with division. It would probably be better to encourage them to use an alternative checking strategy.

A similar approach may be used for checking division but it is probably simpler to add the products of each multiplication, along with the remainder.

Answers

Q. What else could you have done to check the answer?

A. 3,015 ÷ 67 = 45

1. (a) 4,902 ➔ 4,902 ÷ 86 = 57
   or 4,902 ÷ 57 = 86

   (b) 5,135 ➔ 5,135 ÷ 65 = 79
   or 5,135 ÷ 79 = 65

   (c) 7,802 ➔ 7,802 ÷ 83 = 94
   or 7,802 ÷ 94 = 83

   (d) 8,848 ➔ 8,848 ÷ 79 = 112
   or 8,848 ÷ 112 = 79

   (e) 7,102 ➔ 7,102 ÷ 134 = 53
   or 7,102 ÷ 53 = 134

   (f) 7,938 ➔ 7,938 ÷ 147 = 54
   or 7,938 ÷ 54 = 147

   (g) 111,254 ➔ 111,254 ÷ 286 = 389
   or 111,254 ÷ 389 = 286

   (h) 78,540 ➔ 78,540 ÷ 165 = 476
   or 78,540 ÷ 476 = 165

   (i) 171,402 ➔ 171,402 ÷ 294 = 583
   or 171,402 ÷ 583 = 294

2. *Using the inverse operation—multiplication.*

   (a) 72 ➔ 72 x 9 = 648

   (b) 65 ➔ 65 x 8 = 520

   (c) 36 ➔ 36 x 7 = 252

If you are not sure whether the answer to a calculation is correct you can repeat it. Instead of just doing the calculation in the same way you can check multiplication by division and vice versa.

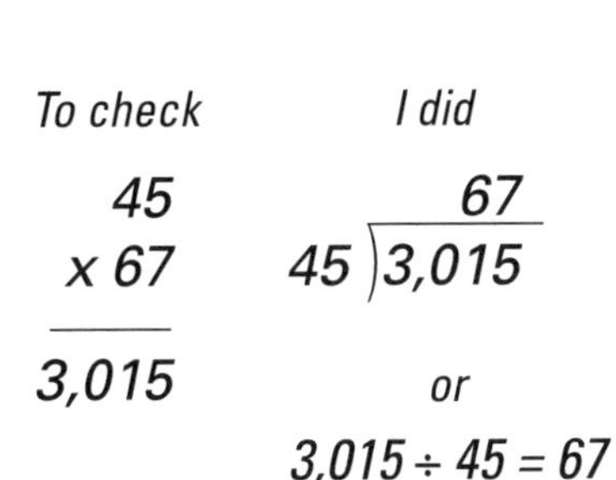

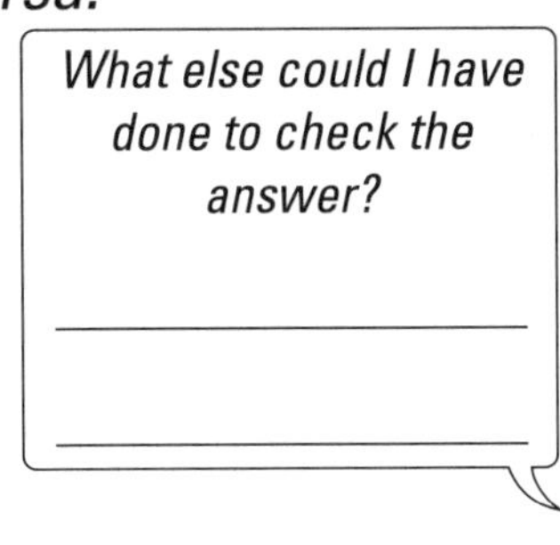

1. Complete the following calculation and then check using an inverse operation.

(a) 57 x 86 = 4,902 — check: 4,902 ÷ 86 = 57 or 4,902 ÷ 57 = 86

(d) 112 x 79

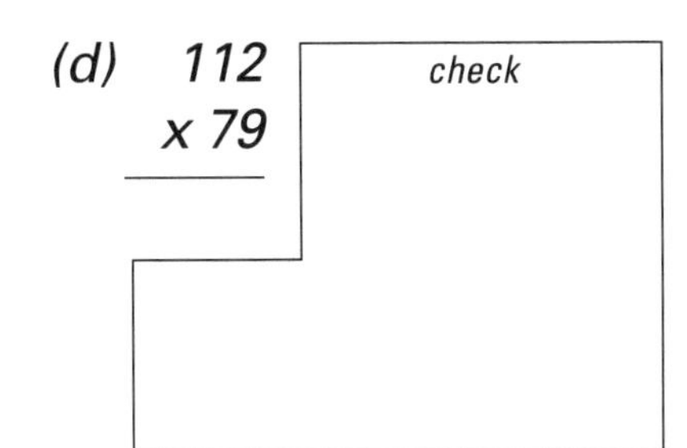

(g) 389 x 286

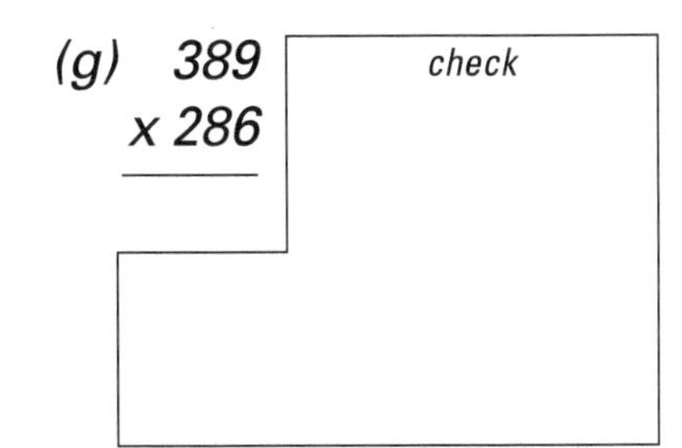

(b) 79 x 65

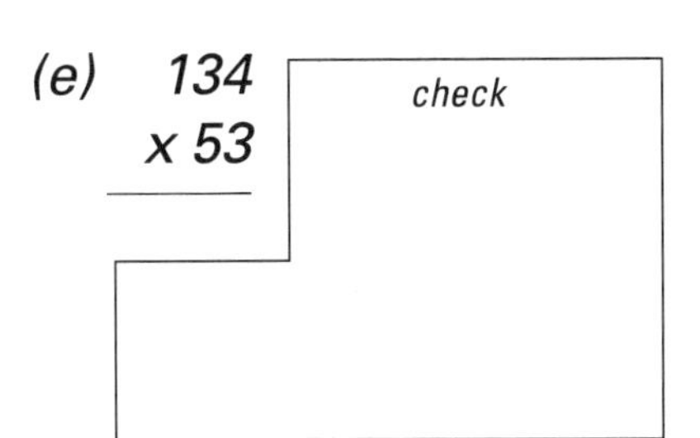

(e) 134 x 53

(h) 476 x 165

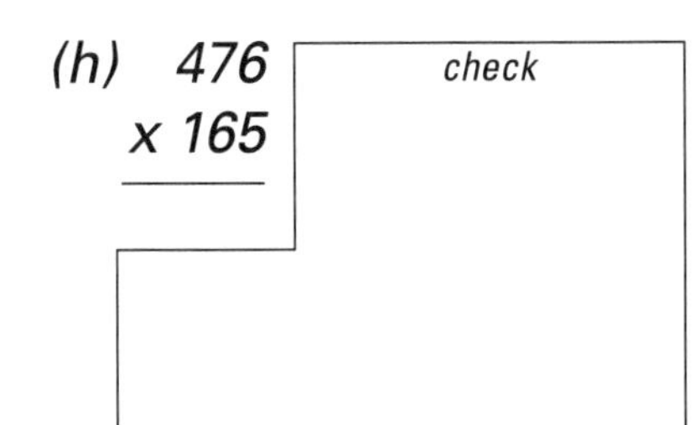

(c) 83 x 94

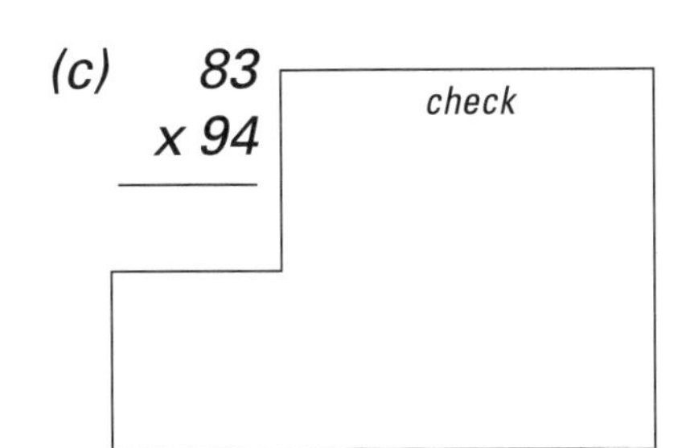

(f) 147 x 54

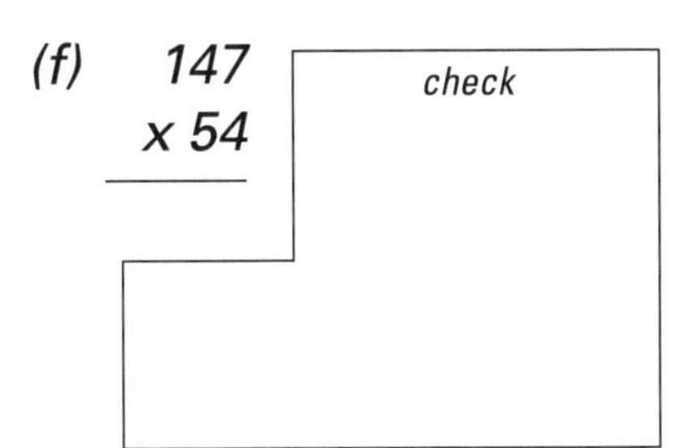

(i) 583 x 294

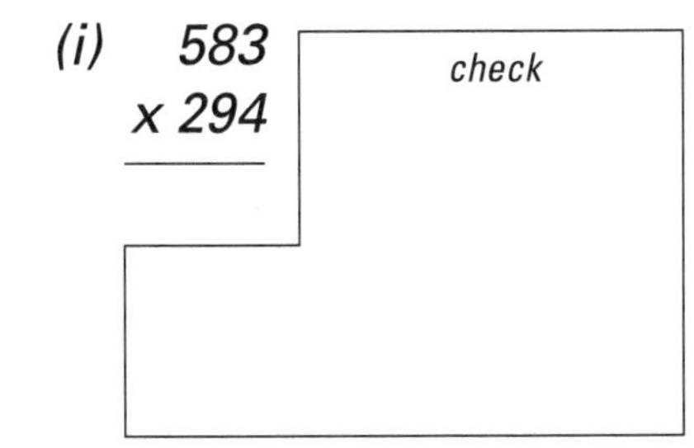

2. Explain how you would check a division.

____

____

(a) 9)648

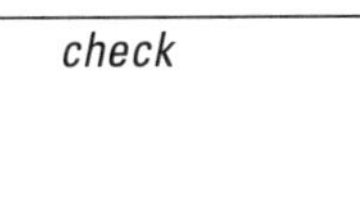

(b) 8)520

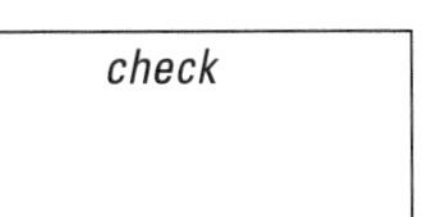

(c) 7)252 — check

Review

Check up – 1

Teacher information

You may use this page to assess student understanding of some of the checking techniques outlined in this book. When marking this page, focus on discussing the reasons for rejecting the answers.

Answers

1. *Odd + odd = even – the answer in this case is odd.*
2. *Even + even = even – the answer in this case is odd.*
3. *Even + odd = odd – the answer in this case is even.*
4. *Even x even = even*
5. *Look at last two digits 2 x 4 = 8 – 248<u>4</u> cannot be right.*
6. *Order of magnitude issue – too large.*
7. *Cannot end in 5.*
8. *7 x 3 = 21– decimal point in the wrong place.*
9. *4 + 2 + 1 ➔ 7 – Digit sum should be a three or a multiple of three.*
10. *1 + 2 + 3 + 2 ➔ 8 – Digit sum should be 9.*

Check up – 1

Explain why you would reject the following answers.

1. 137 + 235 = 373

2. 138 + 234 = 373

3. 138 + 233 = 372

4. 76 x 34 = 2,583

5. 72 x 34 = 2,484

6. 72 x 34 = 22,448

7. 72 x 34 = 2,485

8. 7.2 x 3.4 = 244.8

9. 137 x 3 = 421

10. 137 x 9 = 1,232

Review

Check up – 2

Teacher information

Investigations 1 and 2 focus on the pattern that occurs when a number is multiplied by 5; i.e. the answer will end in 5 or 0.

Investigation 3 involves multiplying by an even number. The result should always be even.

Investigation 4 involves examining the patterns formed by multiplying *odd x odd, odd x even* and *even x even* numbers.

Answers

1. *Answer will end in 5 or 0.*
2. *Answer will end in 5 or 0.*
3. *Result will be even.*
4. *odd x odd = odd*
   odd x even = even
   even x even = even
5. $\begin{array}{r} 951 \\ \times\ 73 \\ \hline 69{,}423 \end{array}$
6. *Teacher check*

Check up – 2

Investigate what happens in each of the following situations. You will need to try some examples on the back of this sheet.

1.

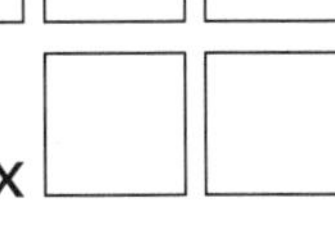

2.

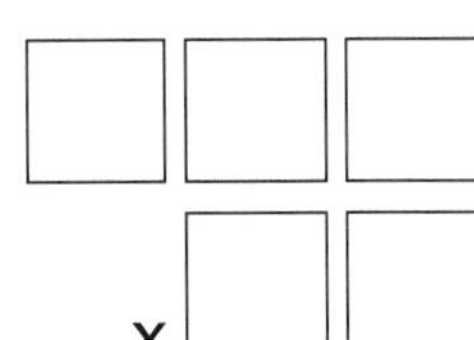

3. Try 2, 4, 6, or 8.

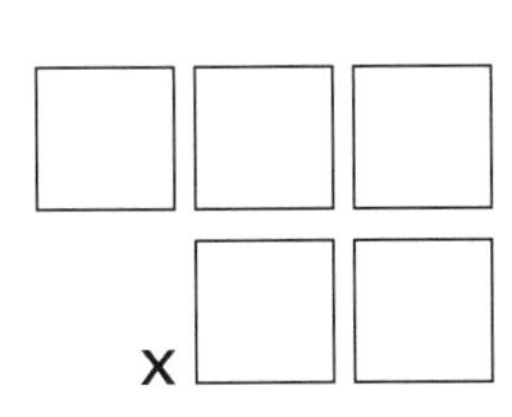

4. Experiment with odd or even digits in these spaces.

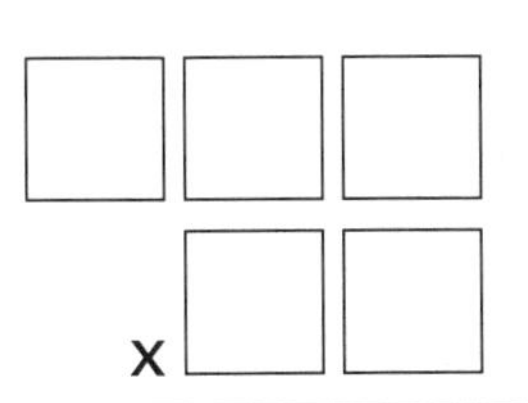

5. Using the digits 1, 3, 5, 7 and 9, try to create the largest answer you can.

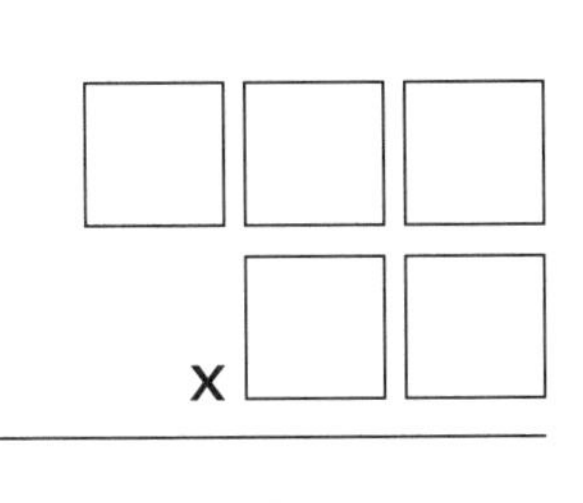

6. Now try using the digits 2, 4, 6, 8 and 0.

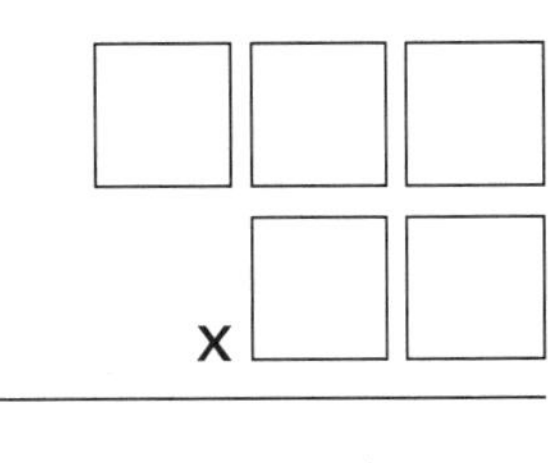

Number charts

0	1	2	3	4	5	6	7	8	9
10	11	12	13	14	15	16	17	18	19
20	21	22	23	24	25	26	27	28	29
30	31	32	33	34	35	36	37	38	39
40	41	42	43	44	45	46	47	48	49
50	51	52	53	54	55	56	57	58	59
60	61	62	63	64	65	66	67	68	69
70	71	72	73	74	75	76	77	78	79
80	81	82	83	84	85	86	87	88	89
90	91	92	93	94	95	96	97	98	99

0	1	2	3	4	5	6	7	8	9
10	11	12	13	14	15	16	17	18	19
20	21	22	23	24	25	26	27	28	29
30	31	32	33	34	35	36	37	38	39
40	41	42	43	44	45	46	47	48	49
50	51	52	53	54	55	56	57	58	59
60	61	62	63	64	65	66	67	68	69
70	71	72	73	74	75	76	77	78	79
80	81	82	83	84	85	86	87	88	89
90	91	92	93	94	95	96	97	98	99

0	1	2	3	4	5	6	7	8	9
10	11	12	13	14	15	16	17	18	19
20	21	22	23	24	25	26	27	28	29
30	31	32	33	34	35	36	37	38	39
40	41	42	43	44	45	46	47	48	49
50	51	52	53	54	55	56	57	58	59
60	61	62	63	64	65	66	67	68	69
70	71	72	73	74	75	76	77	78	79
80	81	82	83	84	85	86	87	88	89
90	91	92	93	94	95	96	97	98	99

0	1	2	3	4	5	6	7	8	9
10	11	12	13	14	15	16	17	18	19
20	21	22	23	24	25	26	27	28	29
30	31	32	33	34	35	36	37	38	39
40	41	42	43	44	45	46	47	48	49
50	51	52	53	54	55	56	57	58	59
60	61	62	63	64	65	66	67	68	69
70	71	72	73	74	75	76	77	78	79
80	81	82	83	84	85	86	87	88	89
90	91	92	93	94	95	96	97	98	99

1	2	3	4	5	6	7	8	9	10
11	12	13	14	15	16	17	18	19	20
21	22	23	24	25	26	27	28	29	30
31	32	33	34	35	36	37	38	39	40
41	42	43	44	45	46	47	48	49	50
51	52	53	54	55	56	57	58	59	60
61	62	63	64	65	66	67	68	69	70
71	72	73	74	75	76	77	78	79	80
81	82	83	84	85	86	87	88	89	90
91	92	93	94	95	96	97	98	99	100

1	2	3	4	5	6	7	8	9	10
11	12	13	14	15	16	17	18	19	20
21	22	23	24	25	26	27	28	29	30
31	32	33	34	35	36	37	38	39	40
41	42	43	44	45	46	47	48	49	50
51	52	53	54	55	56	57	58	59	60
61	62	63	64	65	66	67	68	69	70
71	72	73	74	75	76	77	78	79	80
81	82	83	84	85	86	87	88	89	90
91	92	93	94	95	96	97	98	99	100

Notes